WORKSHOP PROCESSES AND MATERIALS I

P. J. Avard

J. Cross

Medway and Maidstone College of Technology

M

First published 1977 by
THE MACMILLAN PRESS LTD
London and Basingstoke
Associated companies in Delhi Dublin
Hong Kong Johannesburg Lagos Melbourne
New York Singapore and Tokyo

ISBN 0 333 21132 4

Printed in Great Britain by A. Wheaton & Co., Ltd., Exeter

Contents

Preface   VII

PART ONE   1

Introduction   2

Section 1 Safety at Work   3

Hazards in the Workshop   3
The Health and Safety at Work Act   3
Clothing, Hair and Machine Guards   4
Dangers in the Use of Electrical Equipment   6
Stopping Equipment in an Emergency   7
Eye Protection   7
Improving Workshop Conditions   8

Section 2 Materials   10

Ferrous Metals   10
Non-ferrous Metals   12
Bearing Properties of Different Materials   13
Uses and Properties of Plastics   14
Joining Thermoplastics Using Solvent and Welding
   Techniques   16
Problems Associated with Machining Plastics, and the
   Speeds and Feeds Necessary   16
Bending Techniques for Forming Plastics   17
Casting Techniques for Forming Plastics including
   Encapsulating Components for Electrical Purposes   18
Protection of Metals from Atmospheric Corrosion   18
Methods of Painting Metal Surfaces and the Hazards
   inherent in Spraying and Stoving   18
Selecting the Appropriate Protective Treatment for
   Specific Metals   19

PART TWO   21

Introduction to Process Notes   22

Process 1   Centre Lathe, Turning between Centres   23
Process 2   Centre Lathe Turning, Four-jaw Chuck   27
Process 3   Taper Turning (Compound Slide)   31
Process 4   Screw Cutting (Single-point Tool)   35
Process 5   Turning Eccentrics   39
Process 6   Centre Lathe, Taper-turning Attachment   43
Process 7   Centre Lathe, Face Plate   47
Process 8   Shaping a Vee-groove   51
Process 9   Shaping   57
Process 10   Marking-out, Hand Sawing and Filing   61
Process 11   Marking-out   67
Process 12   Spacing Holes on a Pitch Circle   71
Process 13   Marking-out, Drilling and Reaming   75
Process 14   Marking-out, Heat Treatment, Drilling
             and Bending   81
Process 15   Accurate Hole Spacing   87
Process 16   Soft Soldering   93
Process 17   Milling—Producing a Concentric Square
             on a Round Bar   99
Process 18   Milling—Angular Indexing   103
Process 19   Milling a Four-jaw Coupling   109

Preface

This book is intended primarily for students taking the first-year examination of the Technician Education Council (TEC) in Workshop Processes and Materials I.

The format of the book is rather unusual. The first section deals with *safety*, a topic of vital importance in any syllabus. This is followed by a section on *materials*, comparing the properties of common engineering materials, including plastics. This part of the book also contains information on forming and joining plastics materials, as well as protective treatments for various metals. *Hand processes* and *machine tools* (which carry the bulk of the marks in the assessment specification) are dealt with separately.

Part Two of the book takes a different form, being divided into nineteen sections on *processes*; between them, these sections deal with most aspects of the recommendations of the Standard Unit, introducing the student to practical work at the earliest possible stage. The object is to create learning situations for the students, in which the lecturer can provide his own individual teaching.

Each of the processes has been carefully checked, and has been evolved and modified in a classroom/workshop situation. The *process notes* provide a general guide for the work to be undertaken, while the detailed instructions for carrying out the work for each process are written in the 'shorthand English' found in industry—students will find this realistic and easy to assimilate. Sections of the Standard Unit that do not lend themselves to a practical treatment are purposely omitted, enabling lecturers to deal with them in the classroom in their own way, providing variety in the students' work programme. Test questions have been provided after each process.

PART ONE

INTRODUCTION

The object of this first part is to provide the student with a relatively brief summary of the work required in the *safety* and *materials* sections of the Standard Unit. *Plastics* are also covered, including joining and forming techniques.

SECTION 1

Safety at Work

HAZARDS IN THE WORKSHOP

Being aware of the hazards present in an engineering workshop is the first step towards avoiding them.

An enormous amount of expense and expertise has gone into the design of safety features of machine tools and the laying out of efficient workshops, the object being to enable the skilled craftsman to operate more efficiently in the safest, most comfortable conditions possible, bearing in mind the particular nature of the skill being performed.

Any workshop must be considered as a potentially dangerous place where minor and fatal accidents can and must be avoided. Foolishness, carelessness and malpractice must be eliminated from a machine-tools operation and common sense must prevail.

Remember that *machines cannot be made foolproof* and ultimately the working environment is only as safe as the operators within it.

Building up confidence is an essential part of an apprentice's training; concentration on the operation actually 'in hand' is vital.

Personal hygiene should never be neglected; it forms an important part of the self-discipline needed for the competent craftsman; failing to observe basic rules can lead to contracting industrial diseases.

THE HEALTH AND SAFETY AT WORK ACT

The aim of the *Health and Safety at Work Act* is to promote greater safety awareness and to provide stimulation so that all parties concerned fulfil their obligations to participate in raising standards.

A legislative framework has been designed to enforce the provisions of the Act, and contravention can ultimately lead to a fine of up to £400 or, for certain offences, imprisonment for up to two years.

The major responsibilities of the employee under the 1974 Act are laid down in its sections 7 and 8 as follows.

Section 7 a It shall be the duty of every employee while at work to take reasonable care for the health and safety of himself and of other persons who may be affected by his acts or omissions at work; and

b as regards any duty or requirement imposed on his employer or any other person by or under any of the relevant statutory provisions, to co-operate with him so far as is necessary to enable that duty or requirement to be performed or complied with.

Section 8 Duty not to misuse. No person shall intentionally or recklessly interfere with or misuse any thing provided in the interests of health, safety or welfare in pursuance of any of the relevant statutory provisions.

CLOTHING, HAIR AND MACHINE GUARDS

Overalls

A vital part of an apprentice's training is to develop good habits of dress. Unfortunately trends in today's fashion do not take into account industrial safety and the suitability of the style under working conditions!

It is obvious that the degree and type of protective clothing depend on the nature of the work being done. For instance, it would be unreasonable to expect a watchmaker to wear hard hat, goggles, boiler suit and industrial footwear at work; equally it is just as unreasonable for a machine operator to report for work wearing plimsolls and swimming trunks.

To meet most types of conditions, a boiler suit/overall is most serviceable and gives maximum protection. The suit must be close fitting but allow free movement, preferably with short sleeves or at least with the cuffs turned back.

Ties, which should either be removed or tucked away, flared trousers and woolly sweaters all constitute additional hazards when dealing with equipment that has moving parts.

There are two main fabrics from which the boiler suit can be made: heavy-weight cotton drill or nylon. The choice is at the discretion of the wearer, but remember that a nylon boiler suit is not recommended for welding because of the added fire risk.

Photo 1 The horrifying effects of catching long hair in a lathe leadscrew

Leather blacksmith's aprons give extra protection and also prevent holes being burnt in overalls, which are expensive and need looking after.

Whatever the individual preference in overalls, they need proper and regular laundering. This promotes greater personal hygiene and makes you more acceptable to your colleagues who have to work close to you!

Photo 2 A 'Tensaguard' safety cover being fitted to a lathe
(Photograph courtesy of Tensator Limited)

Hair

If ever there was a fashion to be regarded with bitter regret in an engineering workshop it is the long hairstyle adopted by many young people.

During wartime many women were employed in munitions factories and the dangers of long hair were made painfully apparent; it was a common occurrence for women to suffer some accident involving their hair. Even today individuals are slow to appreciate the consequences of entanglement with moving machinery. A range of inexpensive headgear, incorporating a snood, has been developed to meet the safety requirements of people in industry. A snood is a band or net with which hair is tied up and contained in a controlled position away from likely dangers. Eventually it is hoped that this type of headgear will become as common as the hard hat in the building and construction industry.

Apart from the obvious safety factors inherent in wearing headgear, it helps to keep out unwanted grease and oil to aid greater cleanliness.

Reference

Department of Education and Science, D.E.S. Safety Series No. 5, Safety in Further Education: chapter 4 Practical Areas—Machinery.

Footwear (BS 1870: Safety footwear)

The importance of proper footwear and the type of protection it provides are often overlooked.

Many industrial injuries are caused by objects falling on to an operative's feet, which can lead to badly crushed feet or even the loss of the limbs. Plimsolls and soft shoes offer no protection.

Industrial boots provide protection mainly against three types of injury

(1) injury to the toes, by providing steel toecaps
(2) injury from sharp pieces of metal piercing the feet, by reinforced non-slip soles

(3) serious injury to the Achilles tendon, by extending upwards over the ankle; this type of protection is particularly desirable in heavy fabrication shops where stacking of sharp-edged sheet metal and jagged flame-cut components is commonplace.

With ever-increasing costs a point worth bearing in mind is that properly designed industrial footwear gives far better value for money in terms of wear and tear than do everyday shoes.

Guards

It is very easy to subordinate safety by operating machines and equipment without guards. As was pointed out earlier, an enormous amount of money has been spent on the design and production of devices to enable operatives to work as safely as possible.

Sometimes you may find it awkward to carry out a particular machining operation with the guards in place or to view the actual working area. Under no circumstances should guards be removed to accommodate the work unless suitable alternative safety devices are adopted.

The safest method of guarding is by *fixed* guard. This type is permanently fixed to the machine, requires no adjustment and cannot be removed by the operator. *Trip* guards are a valuable contribution to safety; these are often found on a guillotine where metal is fed into cutting blades by hand—the position of the trip mechanism can be adjusted to allow the thickness of the metal to be passed through, while anything thicker, such as fingers, will automatically lift the bar and apply a brake to the machine.

Because of the variation in work, shape and size, guarding the cutters efficiently on a milling machine is very much more complex. There are numerous designs of *milling cutter guard* and almost all rely on the operator's own thoughtfulness to implement them. Spring-loaded *telescopic guards* are an efficient method of preventing accidental contact with a revolving drill or drill chuck. The design completely envelops the moving parts and is common on larger drilling machines. *Chuck guards* on centre lathes stop the operator making contact with the rotating chuck; they are usually hinged to provide access. Transparent guards are particularly useful, enabling the turner to view the work from behind a protective screen.

Spring-loaded telescopic guards are seldom fitted on the lead and feed shafts, but can serve the added purpose of protecting the components from damage and accumulation of dirt.

As a constant reminder that guards are of particular importance they should be painted in an outrageously bright colour contrasting with the machine; when not in use they should be stored in an easily accessible place.

References

Department of Education and Science, D.E.S. Safety Series No. 5, Safety in Further Education: chapter 5 Workshops.
Health and Safety at Work, Booklet No. 20, Drilling Machines—Guarding of Spindles and Attachments.
Health and Safety at Work, Booklet No. 33, Safety in the Use of Guillotines and Shears.
Health and Safety at Work, Booklet No. 42, Guarding of Cutters of Horizontal Milling Machines.

DANGERS IN THE USE OF ELECTRICAL EQUIPMENT

Electricity is particularly hazardous in a machine shop, where there are numerous metallic objects ideally situated to expose personnel to a serious electric shock.

Exposed unprotected cables can become damaged through normal workshop routine and lead to lethal consequences. The wiring and maintenance of electrical equipment is a skilled task and should be carried out by qualified electricians.

(1)   Water and coolants should be kept away from all electrical apparatus.
(2)   Cable runs for portable hand tools should be up and over gangways (above head height), stopping people tripping over the cable and keeping it away from normal gangway traffic. If cables must be laid in the workshop gangways, they should be protected by ramps and suitable warning notices displayed.

(3)   Worn or chafed lengths of cable with little insulation should only be repaired with tape as an immediate temporary measure by the electrical maintenance department.

(4)   No electric lamp should be used without a shade.

(5)   Power tools should only be connected to correctly installed power points and not to light sockets. Light sockets have no means of conducting the earth connection to the power tool.
Light sockets are for electric lamps only

(6)   All machine lighting should be low voltage. Earth faults can develop through the exposed light flex becoming chafed, or the connections being immersed in cooling liquid.

(7)   All portable lamps should be low voltage.

Reference

Department of the Environment Advisory Leaflets, No. 18 Powered Hand Tools 1: Electric Tools; No. 20 Powered Hand Tools 3: Safety and Maintenance.
Department of Education and Science, D.E.S. Safety Series No. 5, Safety in Further Education: chapter 3 Precautions in the Use of Mains Electricity, chapter 4 Practical Areas—Electrically Powered Hand Tools.

STOPPING EQUIPMENT IN AN EMERGENCY

Within the workshop layout there must be a system that can easily isolate the entire machine shop or any individual machine. To be effective the positioning of the stopping arrangement in relation to the machines and operators is of the utmost importance.

Equipment can be stopped in an emergency either by mechanical means or by electrical means.

(1)   Mechanical methods of stopping individual machines are normally simple and effective. Quite often a single lever will start and stop the machine. On normal applications stopping is performed gently and the spindle comes to rest slowly. This lever mechanism can also incorporate a break where more severe operation arrests the main spindle, stopping it immediately.

Colour coding is used to distinguish the stop/start lever from other controls and normally has a contrasting red knob for easy identification. To supplement the normal stop/start arrangements, a foot-operated treadle is favoured by manufacturers as a sure, positive type of emergency stop. This would be of particular value if an operator's hands were to be immobilised through entanglement with the moving parts of the machine.

(2)   Safety is a major factor to consider when designing both mechanical and electrical apparatus. The positioning and effectiveness of electrical switchgear are often overlooked. Most modern machines are equipped with switches that have similar characteristics, allowing easy access to stop buttons and more difficult access to start buttons. Both are positioned on the same panel within arm's reach of the normal operating position. Stop switches are usually large bulbous buttons coloured red and are designed to stand out above the facia of the panel. Ideally the stop control should only perform the single function of stopping, not stopping and then reversing, since this defeats the object of a positive stop and its effectiveness as a brake. On the other hand, start buttons are recessed or shrouded so that the possibility of accidentally starting the machine is considerably reduced.

Situated in obvious and accessible positions around the workshop should be a series of *emergency stop buttons* any one of which, when pressed, will immediately switch off the electrical supply to the entire machine shop. This does not apply a brake to stop each machine immediately nor does the switch affect general workshop lighting. Access to these buttons should never be blocked by careless stacking of materials, and notices displaying the presence of the buttons must be displayed clearly.

EYE PROTECTION

BS 2092: Industrial Eye Protectors

Injuries to the eyes happen all too frequently. Because of the regularity and nature of damage to eyesight, great emphasis should be given to wearing eye protection when there is the remotest chance of flying particles.

Often the reason why an operator does not bother to wear goggles or the eye protection supplied is that they have been poorly maintained and have become unfit for use. For instance, a transparent shield fitted to a machine will, if not regularly cleaned, become opaque, and in due course the operator will remove it so that he can see the machining area.

Some employers operate a scheme whereby they issue eye protection to their employees on a permanent loan basis. This ensures that the operator keeps the goggles in a good state of repair, and also reduces the risk of contagious infections being spread. Employers operating this scheme must be sure of selecting, from the wide choice available, the correct type of protection for the job on which each machinist is engaged.

Perhaps one of the most common sources of injury is the short grinding job on an off-hand or surface grinder. Abrasive particles leave the wheel at speeds approaching that of a rifle bullet. All too often some are deflected into the face and perhaps the eyes of the operator. It is the speed at which they travel, and the minute size of the particles that make them so dangerous.

Some operations are obviously more hazardous than others and some materials behave differently from others when being machined. An interrupted cut on a centre lathe, machining non-ferrous metals (particularly brass), removing metal using a hammer and cold chisel, chipping slag from an electric weld—are only some of the typical workshop processes that take place in normal everyday routine and where eye protection is vital.

Remember that you can walk to work on a false leg, or even lift a pint of beer with a false arm, but you can never see to read your favourite magazine with a glass eye.

Reference

Department of Education and Science, D.E.S. Safety Series No. 5, Safety in Further Education: chapter 4 Practical Areas—Protective Clothing and Equipment.

IMPROVING WORKSHOP CONDITIONS

It is fair to assume that a tidy well-kept workshop is a safer place to work in than one that is poorly kept and untidy. So in seeking to prevent accidents we must look carefully at our working conditions and, as a start to solving the problem, make sure they are as good as possible.

All too often poor factory layout and bad housekeeping are part of a combination of events leading to accidents that need not have occurred if a conscientious approach by management and employees had been adopted.

Unfortunately it is not difficult to find workshops in need of improvement. Lighting is often well below standard and lacking maintenance. Floor areas are very often untidy, dirty and caked in a thick compound of swarf, grease and oil; gangways are unmarked, choked and clogged with disorganised piles of work; stairways are cluttered with odd rubbish; tools and materials are often poorly stored among swarf; used oily wiper rags are left in piles, creating a fire hazard; bench tops may be fully laden with tools and equipment jumbled together haphazardly. A factory in this sort of condition quite often has a high accident record and an unnecessary number of working hours are lost through injury. The end-product suffers because the workshop cannot operate efficiently and workers are unable to give full attention to the job, thus jeopardising the quality of the product.

Improving these conditions will not involve an expensive programme, just effort, largely on the part of individuals. Benches can be kept tidy. Wiper rags, scrap and swarf can be placed immediately into their correct receptacles; bins for this purpose should be easily identifiable and in plentiful supply. By a coordinated effort between workers and managers gangways can be marked and kept clear. Once these relatively easy-to-clear hazards have been eliminated, the task of keeping the workshop in a more acceptable condition will be made easier.

Thus the key to safer conditions is 'good housekeeping', and management can often take the lead in meeting these needs. Some really conscientious managers operate schemes whereby a bonus is paid to the section with the lowest accident record over a set period of, say, a working week or month. Alternatively, a weekly intersection competition is held and judged by a selected panel, with points awarded to the best-kept section. The department with the highest total over a specified period is considered the winner. The members of that section are awarded some form of bonus, usually financial. Rewards create a friendly rivalry and interest between sections, working wonders for morale and thus ultimately improving output efficiency and accident records.

SECTION 2

Materials

Materials used in the engineering industry can be divided into three categories: (1) ferrous metals, (2) non-ferrous metals and (3) non-metallic materials.

FERROUS METALS

These metals are based on iron and are magnetic.

Wrought Iron

Wrought iron is a very fibrous metal, it is soft, malleable and forgeable but little used in industry today. Its place has been taken by some of the lower-strength steels, which are easier and cheaper to make.

Cast Irons

Grey Cast Iron

This universally used metal has a dark grey colour and a fine porous structure, which can be seen on its machined surfaces. When molten it readily flows into intricate moulds, producing the finished shape of the required work and leaving a machining allowance where necessary. It has low shrinkage values and remains stable under changing temperature conditions. It is the traditional material for machine-tool frames, slide-ways, large-surface tables and heavy instrument structures.

Form of supply: castings
Properties: soft, brittle, easily machinable, low tensile strength but good compressive values

White Cast Iron

This is similar to grey cast iron in appearance, but fractures show a whiter and more crystalline surface; otherwise it has similar characteristics except that it is harder and more brittle. It has limited uses, which are confined to circumstances where hard-wearing surfaces are required. Chills are often placed in the moulds to increase these effects.

Form of supply: castings
Properties: very hard and brittle; more difficult to machine than grey cast iron; low tensile strength, good in compression

Malleable Cast Iron

This has the outward appearance of plain cast iron but by a rearrangement of its carbon structure it has properties approaching that of the low-strength steels. Malleable castings are used in many industrial applications where strength, ductility, machinability and resistance to shock are important factors. The bench vice is a good example of the shock and loads that a good malleable casting will withstand.

Form of supply: castings
Properties: improved tensile strength, malleable to a degree, less brittle and slightly harder than grey cast iron, has good compressive strength

Steels

Low-carbon Steel (Mild Steel)

This steel has a universal use in the engineering industry. The important factor governing its properties is its carbon content, which, although amounting to a very small percentage of its whole structure, has great influence on its strength and adaptability. It can be forged, rolled, drawn and pressure formed. It is not greatly affected by heat treatment except in case hardening, when additional carbon is introduced into the outer surfaces providing a hard case with a soft core capable of withstanding shock loads and at the same time providing a hard-wearing surface.

Form of supply: rolled and drawn bar in many different shapes; sheet, strip and wire; forgings
Properties: malleable, ductile, comparatively soft and tough, with good tensile strength

Medium-carbon Steel

This is steel with improved mechanical properties over low-carbon steel and is used for strong forgings such as crankshafts, springs, tipped and butt-welded tool shanks and reasonably strong castings. Its strength and toughness can be improved by heat treatment but not to the extent of using it for metal-cutting tools.

Form of supply: similar to low-carbon steels
Properties: tough, harder and less malleable than low-carbon steel

High-carbon Steel (Tool Steel)

This is a steel that is tough and hard in its normal condition and responds to heat treatment in a very adaptable manner. It can be made extremely hard, but then it becomes rather brittle. This hardness can be decreased by tempering to bring about a desired degree of hardness combined with toughness—a condition required by many types of cutting tool (such as chisels). Silver steel is another form of high-carbon steel, which is often used in workshops for making special small tools.

Form of supply: castings, hot rolled bar and forgings
Properties: tough, hard, with high tensile strength

High-speed Steel

This is a special steel outside the range of plain carbon steels. It has the property of retaining its hardness and strength at dull red temperatures that occur during machining, and for this purpose it has been developed for cutting tools. A typical composition is carbon 0.6 per cent, tungsten 18 per cent, chromium 4 per cent, molybdenum 1 per cent, vanadium 1 per cent, cobalt 7 per cent, the remainder being iron.

Form of supply: forged bar and block forms, small ground square sections
Properties: hard, tough and strong, characteristics that it retains at dull red temperatures

Composition of Ferrous Metals

Metal	Carbon	Silicon	Phosphorus	Manganese	Sulphur	Density (kg/m^3)	Melting point $(°C)$	Electrical conductivity $(S/m \times 10^6)$
Wrought iron	0.2	0.1	0.06	0.10	0.02	7700	1540	9.3
Grey cast iron	3.5	2.5	1.50	1.00	0.02	7200	1150	1.4
White cast iron	4.0	1.5	2.00	0.50	1.00	7300	1150	1.0
Malleable cast iron	1.5	0.7	1.50	0.25	0.08	8000	1150	3.4
Low-carbon steel	0.25	0.3	0.04	0.40	0.05	7800	1450	5.0
Medium-carbon steel	0.65	0.3	0.04	0.60	0.05	7800	1450	8.3
High-carbon steel	1.50	0.3	0.04	0.70	0.05	7800	1450	8.5

Balance in iron

NON-FERROUS METALS

Aluminium

This is a non-magnetic metal with a silver-white colour, light in weight and, as a pure metal, has low strength values. Usually it is manufactured as an alloy with small percentages of other elements, which greatly improve its mechanical properties. It responds to most forming operations such as rolling, drawing and extrusion, operations that improve its strength. Because of its low weight-to-strength ratio it makes a large contribution to the aircraft and transport industry. It is non-corrosive in normal atmospheric conditions.

Form of supply: plate, sheet, foil, bars, tubes and castings
Properties: soft, malleable, ductile and strong in alloy form

Duralumin

This is a well-known aluminium alloy with a strength value comparable with some steels but at one-third of the weight. By slight changes in its composition for differing processes it can be forged, cast, rolled, drawn and extruded. It has the distinctive property of age hardening and for this reason needs to be annealed before cold working. It is used to a great extent in the aircraft industry but increasingly also in the domestic appliance field.

Form of supply: similar to those for aluminium
Properties: harder and stronger than aluminium

Copper

This is a valuable pure metal with many uses in industry. It is easily identified by its red colour and by its malleable and ductile properties. It is very easy to work by hand and although it work hardens after a while its normal workable condition is rapidly restored by heat treatment. It is extruded, rolled and drawn into many shaped sections especially in the electrical industry, where it finds many uses because of its excellent conductivity. It can be welded, brazed and soldered.

Form of supply: wire, rods, bars, plate, sheet and foil
Properties: soft, malleable and ductile; a good conductor of heat and electricity

Brass

Brass is a copper–zinc alloy. There are several types, each has been developed for its particular property in some manufacturing process, the main difference being in the ratio of copper to zinc. A common brass has a ratio of 70 per cent copper, 30 per cent zinc; this is a strong ductile material suitable for cold working, rolling, drawing and deep pressing operations.

Form of supply: sheet, strip, foil, rods, bars
Properties: high tensile strength, ductile and comparatively hard

Another well-known brass has 60 per cent copper, and 40 per cent zinc. This alloy is best suited for hot forming operations; it is slightly more brittle and not as strong as the 70–30 brass.

Form of supply: formed bar, hot forgings and pressings
Properties: soft, moderate malleability and medium tensile
    strength

Bronze

Phosphor Bronze

This is an alloy of copper, tin and phosphorus. It is a strong, hard and tough alloy, resistant to corrosion and shock loads. It is a good castable material, often taking the place of cast iron where the strength gained outweighs the additional cost. It is used for bearings, gears, springs and filtering screens and for many other engineering components. Its demand as a bearing metal is such that metal suppliers have found it an advantage to stock short cast and machined tubes—called quills—of standard sizes with a machining allowance on the outside and inside diameters for final machining to requirements at the purchaser's own workshop.

Form of supply: castings, cored tubes, solid bar, wire and
    spring strip
Properties: hard, tough, high tensile strength, machinable

Gun Metal

This is an alloy of copper, zinc and lead—a very strong non-corrosive metal, used a great deal in marine engineering. Owing to its strength and toughness it is used for valves and cocks in high-pressure steam systems and for pumps and mechanisms subject to the corrosive effects of sea water.

Form of supply: castings and wrought bar
Properties: high tensile strength, tough and hard, machinable

Magnesium Alloys

Magnesium alloys have been developed mainly for the aircraft industry, since they are lighter than aluminium alloys and approximately equal in strength. Production is beset by many problems, which make magnesium more expensive than aluminium, but the production difficulties are slowly being overcome and already the motor vehicle industry is able to use more magnesium products, the higher costs being offset by the improvements gained. Its mechanical properties are similar to aluminium with some improvement in hardness. It is a good conductor of heat and electricity and is non-magnetic, but is subject to corrosion by salt solutions. The alloys cast well and in this form are used for aircraft spar junctions, landing-gear fittings, in internal weight-saving items like seat frames and in door structures. In ground transport, they are used for car and motorcycle wheels. Considerable care is necessary in machining, since metal dust and chippings can burst into flames owing to overheating.

Form of supply: castings, extruded sections, rolled bars and
    plate
Properties: good tensile strength (especially in extruded sec-
    tions), moderate hardness, brittle

BEARING PROPERTIES OF DIFFERENT MATERIALS

Cast Iron

Cast iron has been used for bearings for many years, but is rapidly losing ground to other materials. For journals of slowly rotating shafts or for sliding surfaces it is still the most economical material. It has the capacity for absorbing vibration and will continue to function reasonably well for a while even when the bearing has been allowed to run dry by neglect—the free graphite in the cast iron providing its own lubricant. Cast iron bearings are usually seen to be rather long in comparison with the shaft diameter; this is to spread the load over the maximum area.

Phosphor Bronze

This is a very popular bearing material, which, when used with hardened steel shafts, has a low coefficient of friction. Lubrication is important for running efficiency and these bearings are usually provided with oilways, which are drip or pressure fed. Phosphor bronze is more expensive than cast iron but the bearings are usually in the form of thin shells and do not require the same amount of material as cast iron, so the cost difference is not great, and is compensated for by the improved efficiency and longer life.

White Metal

White metal bearings have a long proven efficiency in their application to the internal combustion engine. There is no better example of its ability to withstand violently changing loads at high speeds than its use for connecting-rod big-end bearings on a crankshaft. Good lubrication plays an important part in its successful functioning, providing a film between the sliding parts and carrying away the heat generated by friction. The bearing consists of a steel or phosphor bronze shell internally lined with a thin layer of white metal, bonded to the shell. The assembly is supported in a firm housing. White metal is an alloy of tin, copper and antimony and has very low frictional properties as a bearing for steel shafts.

Nylon

Nylon is one of the new bearing materials, which is becoming increasingly used in place of phosphor bronze and cast iron. It has many advantages, such as low frictional values and self-lubrication; it is mechanically strong, hard and tough. At present it is confined to the smaller sizes of bearing of medium loads, where servicing is infrequent or the bearing difficult to reach. Since its melting point is about 260 °C it cannot be used in conditions of high temperature.

Polytetrafluoroethylene (PTFE)

PTFE is a plastics not unlike nylon in appearance but with greatly improved mechanical properties as a bearing material over a wide range of temperature conditions not exceeding 260 °C. It is self-lubricating, dimensionally stable, harder and stronger than nylon

but more costly than any of the materials listed above. It does not mould readily like nylon but has to be preformed as a powder by pressure and then sintered in a baking process, which adds to its cost. It has been successfully used for quite large bearings running at medium speeds and loads.

Graphite

This is another type of composite bearing, which allows for long periods between servicing. Originally they were made of a mixture of phosphor bronze powder and graphite, formed into bearing sleeves under high pressure and heat treated to constitute a porous alloy. They are frequently used for motor vehicle starter motors and generator bearings. Recent developments have now produced a mixture of low friction plastics and metal powders formed and sintered in the same way to provide bearings in machine plant used in the food and fabric industries, where lubricants might contaminate or damage the product.

USES AND PROPERTIES OF PLASTICS

Thermosetting Plastics

Phenol Formaldehyde

This is one of the oldest thermosetting plastics and one of the least expensive. There is a variety of types with slightly differing properties depending on the filler material used. Generally speaking they are hard and tough and have good chemical and electrical resistance but are inclined to be brittle in some of the filler combinations. Machining is difficult, the material having severe abrasive and clogging action on the tool. Most of the products are made by moulding and these range from fans, distributor covers, filler caps, switchgear casings, steering wheels, large micrometer hand heat shields, slip gauge boxes and machine tool inspection covers. A well-known fabric laminate is *Tufnol*, which is machinable and is used for gears and electrical instrument panelling.

Urea Formaldehyde

This is similar to phenol formaldehyde but with a harder surface and an unlimited range of colouring, which makes it suitable for decorative panels, artwork in domestic items, and electrical fittings. It is also used in paints and lacquers to give a hard bright gloss. Strong acids affect the finished surfaces and its resistance to water and heat is lower than that of phenol formaldehyde.

Polyester Resin

This is used as the bonding part of a plastics, made up of layers of glass fibre, fabrics or fine metal mesh, which is finally hardened and toughened by a chemical hardening agent reacting with the resin. The required thickness and strength are built up by increasing the number of layers. In this way, for example, the shell of a boat hull can be built on a mould of the required shape either by hand layering or spraying; the shell thus obtained is strong, flexible and light. Car bodies, motorcycle fairings, electrical switch panels, furniture, corrosion-resistant liquid containers and hollow streetlight standards are built up by similar methods. Thus it can be seen to be a versatile material, tough, flexible and strong.

Thermoplastics

Cellulose Nitrate (Celluloid)

This is a clear transparent highly inflammable material. It was an early development in the plastics field and is slowly being displaced by safer materials. Where there is no danger of combustion it is still used for metal coating for decoration or weather protection, small inspection windows, eye-shield frames and many applications of art finishes on domestic items. It is reasonably tough, flexible and receptive to colouring.

Cellulose Acetate

This is very similar to cellulose nitrate but with a slow combustion rate. It is used for machine tool swarf shields and splash guards, control handle knobs, small motor vehicle panels, tool handles and films. It is tough, with a good impact strength.

Acrylics

This is a well-known range of plastics often identified by trade names such as *Perspex* and *Plexiglas*. In one form it has useful optical properties and in other (opaque) forms it is resistant to corrosive atmospheres and the effect of sunlight. It is tough, strong and hard and is used for pressurised windows and aircraft canopies, decorative car panels and small mechanical gear housings. It is available in sheets, rods and castings for local processing such as drilling, tapping and other forms of machining.

Styrene Resin

This is used for electrical insulation, refrigerator panels and trays, food containers, wall tiles and electrical display screens. It is stiff and rather brittle, hard and readily moulded.

Polyvinyl

This is used for pipes, tubes and small containers for water, petrol and hydraulic brake fluids, low-voltage insulation, guttering, floor tiles, wall beadings, corrugated sheeting and building trade fittings. The mechanical properties vary with the plasticiser and filler material used. In general it is hard, tough and self-extinguishing.

Nylon

This is one of the best-known plastics, probably because of its many uses throughout the whole of the manufacturing industry for silent-running gears, minor control mechanisms subject to abrasive wear, bearings, low-stress cam operations, door catches, hinges, tubing, domestic containers and even hard-wearing brush bristles. Mechanical properties are: self-lubricating, high tensile strength, corrosion resistance and resistance to fatigue stresses.

Polythene

This is a comparatively cheap product with many uses, such as cable insulation, tubing, coloured sleeving on electrical leads, transparent roofing, etc. It has endless applications in the domestic field for bowls, buckets and miscellaneous containers. It responds readily to the many forming processes such as moulding, extruding and rolling. Electrical resistance is very good, it is moderately hard and flexible but is not self-extinguishing in combustion.

Polytetrafluoroethylene (PTFE)

This is one of the more recent developments in plastics, which has proved to be a valuable contribution to mechanical engineering. Unfortunately it is rather expensive, which limits its present use. Like nylon, it is self-lubricating and makes excellent bearings. It is tough, hard and strong yet machines easily and for this reason it can be obtained in standard forms of supply such as sheet, round bar and strip for further processing in general workshops. It has a wide operational temperature range up to a maximum of about 260 °C and is non-inflammable. The range of applications includes silent gear trains, slide facings, cams and medium-size mechanisms subject to high frictional loads. It is perhaps best known as a non-stick lining for domestic utensils.

JOINING THERMOPLASTICS USING SOLVENT AND WELDING TECHNIQUES

In engineering it may often be more acceptable and less costly to produce a component by fabricating it from previously prepared material. A typical example of this type of construction is an oil drilling platform. Similarly, if a plastics component cannot be produced to final shape in one moulding operation it may be built up or *fabricated* by joining other moulded items together. Two methods for fastening thermoplastics together are: (1) solvent or cementing and (2) welding.

(1) For *cementing* thermoplastic materials the surfaces to be joined are soaked in solvent, which softens and melts the plastics.

In this way the plastics actually forms its own adhesive and on contact the two pieces will fuse together. Maximum strength from the bond is obtained by the use of heat to assist in the evaporation of the solvent.

Some adhesives are referred to as *dope cements* and are made up of a solution of the plastics to be joined and a solvent. This method of bonding is more commonly used for assembling badly fitting parts. The cement acts rather like a filler, enabling the joint to be built up, making a stronger, more acceptable join.

(2) Thermoplastics can be *heat welded* in a manner similar to that used for the welding of metals. A hot jet of gas, normally air, is directed through a torch on to the surfaces of the work. Working temperatures must be high enough to allow the material to soften and melt, but too high a gas temperature will cause structural deterioration in the plastics. The *filler rod* is normally of the same composition as the plastics being joined and is available in different diameters to suit the size of work. In *heated tool welding* parts to be joined are heated by electrically heated resistance strips or with a hot plate. The heated parts are held together under pressure and allowed to set. An example of this method of joining can be seen in the manufacture of polythene bags.

PROBLEMS ASSOCIATED WITH MACHINING PLASTICS AND THE SPEEDS AND FEEDS NECESSARY

As we have seen plastics can be divided into two main groups: (1) thermosetting plastics and (2) thermoplastics.

(1) These materials appear to be relatively hard and brittle, behaving in a similar fashion to brass when being machined. An intermittent shower of swarf is produced requiring *eye protection* for the operator. Small rake angles are used on cutting tools, hence twist drills with slow helix angles are preferred. Best results from machining operations are obtained by taking deep cuts with fine feeds, but since these materials tend to be very abrasive, wear at the cutting edge can be a problem. Machining is usually carried out dry or with an air blast. Unlike thermoplastics these materials will not melt and then take up their original form on cooling.

(2) Within this group the materials are generally softer and less

rigid, requiring slightly different machining techniques. Tool geometry needs to be changed with rake angles slightly increased, and twist drills should have a quicker helix than is used for thermosetting plastics. Swarf is produced in a continuous ribbon and for certain materials soluble oil should be in plentiful supply. A characteristic property of thermoplastics is that they melt when heated to fairly low temperatures and when cool return to their original condition. Therefore machining temperatures must be kept as low as possible to prevent undesirable distortion.

The following chart will serve as a useful starting point when selecting speeds and feeds, but because of the large number of plastics containing such a range of properties it is difficult to pinpoint definite cutting speeds. Generally it can be accepted that conditions similar to those for brass will apply. However, more efficient stock removal can be obtained by using cutting tools with edges specifically designed for plastics.

Manufacturers will readily advise on their recommended speeds and feeds to give optimum results.

Material	Surface Speed (m/min)	Feed (mm/rev)	
		roughing	finishing
Thermosetting plastics	30–61	0.075–0.125	0.050–0.100
Thermoplastics	61–91	0.075–0.125	0.050–0.100

BENDING TECHNIQUES FOR FORMING PLASTICS

When heated, thermoplastic sheet can be mechanically pressed, blown or sucked on to the contour of a mould, which may either be male (convex) or female (concave). During the cooling period the forming force remains applied until the plastics become rigid. The formed shape will then be retained when removed from the mould.

The process of thermoforming lends itself ideally to producing relatively simple shapes from thermoplastic sheets (cellulose acetate, vinyl polymer, impact strength polystyrene).

Vacuum Forming

In this process the thermoplastic sheet is heated to the softening temperature in a gas or electrically heated oven. It is then transferred to the moulding machine and secured over the female mould cavity. A vacuum is created between the hot sheet and the sides of the mould through small suction holes. Air pressure forces the pliable plastics against the sides of the mould until the temperature has dropped sufficiently to release the vacuum and remove the work.

A variation on this technique is for producing deep articles such as buckets. The hot sheet is draped over a male former and suction applied through a series of holes round the outside of the former. More complex shapes may include depressions and relieved sections.

Vacuum forming is used to produce suitcase shells, advertising displays, tea trays, etc.

Pressure Forming

Complicated shapes and thicker plastics sheets may require more pressure to form the shape than can be supplied by vacuum. Additional pressure can be supplied by applying compressed air at up to $10.3 \times 10^5/\text{N/m}^2$ (150 p.s.i). Mechanical means may also be used to assist in pushing the plastics into recesses.

Mechanical Forming

Sheets heated to the proper softening temperature can be pressed between matching dies. This can leave surplus material, which needs to be trimmed, then the two halves are joined with adhesive. The method is often used to manufacture hollow toys and babies' rattles.

Sometimes the softened plastics sheet is laid on a contoured wooden forming block. A framework is lowered on to the sheet, which pushes it into shape. Compressed air is then directed on to the surface to cool it before removal from the mould.

CASTING TECHNIQUES FOR FORMING PLASTICS INCLUDING ENCAPSULATING COMPONENTS FOR ELECTRICAL PURPOSES

With the many plastics in use today, and further research into developing new materials continuing, techniques must be adapted and equipment adjusted to suit the new materials.

In *casting* the liquid plastics is poured into a prepared cavity (the mould) and is allowed to set or harden; this is called polymerisation. Certain plastics need to remain in the mould until hardening is complete, but with others components can be removed before the complete setting process has taken place and the final stages of polymerisation allowed to continue at room temperature. Heat is generated during the polymerisation process, and to avoid the adverse effects of overheating it may be necessary to cool the mould. Both thermoplastics and thermosetting plastics can be cast.

The casting process is preferred where small quantities of work are required, and the manufacture of expensive injection or compression moulds would prove uneconomical. It is also preferred when producing simple sectional shapes such as tubes, flat sheets and round bar.

There are three variations of the casting process: (1) embedding, (2) encapsulating and (3) potting.

(1) *Embedding* involves the surrounding of work in a transparent plastics. The technique is ideally suited for mounting small specimens prior to micro-examination, where handling and identification difficulties are considerably eased.

(2) *Encapsulating* involves the work being securely set in firm cellular structured plastics or a plastics foam. The process is often used to protect fragile electrical parts; expensive projector lamps can be packaged in this way to reduce accidental damage during transit.

(3) *Potting* is similar to encapsulating but electrical components are covered in a more rigid plastics, mainly for insulation and protection purposes.

PROTECTION OF METALS FROM ATMOSPHERIC CORROSION

With the enormous variety of engineering metals covering a wide range of applications, an equally comprehensive selection of surface treatments must be made available.

The problem of corrosion with metallic materials varies with the conditions in which they operate. In certain climatic conditions they revert much more quickly to their original non-metallic state and in other conditions this change is more gradual. If the working life of components is to be maximised then the change must be prevented or at least slowed down. For example, if mild steel is subjected to a damp atmosphere the iron content of the steel reacts with the oxygen in the air forming an oxide—rust. The higher the moisture content the quicker the reaction takes place, and surface deterioration occurs much more rapidly in warm, damp conditions.

To be effective, surface treatments must exclude the atmosphere and place the component within a protective envelope without affecting the functional properties of the component.

METHODS OF PAINTING METAL SURFACES AND THE HAZARDS INHERENT IN SPRAYING AND STOVING

Perhaps one of the most extensively used methods of protecting metallic surfaces from corrosion is painting. The process is very convenient and lends itself to a very wide range of applications. Examples of these are motor vehicle bodies, which are often painted mainly for protection purposes, but also to give a more decorative appearance; the hull and superstructure of a ship are painted with lead-based paint to protect the metal from salt-water corrosion.

Method	Process
Barrel enamelling	Small components rotated in barrel containing measured quantity of paint—barrel heated to assist drying—parts removed when 'tacky'—placed in a drying oven.
Dip coating	Suitable for large or small work—dipped into bath of epoxy paint or high-build-up cellulose—can either be dried naturally or in an oven.
Electropainting	Principle similar to electroplating, useful for large awkward shapes—box sections in motorcar chassis—can be fully automated and thickness of paint controlled—no runs or sags—little wasted paint.
Flow coating	Products that can be dip coated can be flow coated but flow-coated parts cannot always be dipped—paint pumped through pipes and out of nozzles above work—allowed to run down over work—surplus paint collected, strained and passed back to tank.
Roller coating	Ideal for flat metal surfaces—work can be painted both sides simultaneously—thickness varied by pressure—very fast—painting of venetian blinds is a typical application.
Spray painting	Widely used—any shape or size of work—quick—large quantity of paint wasted—cellulose and other special rapid-drying paints.

Because of the very inflammable nature of the organic solvents used in the make up of paints, a high potential fire hazard exists. The most common solvent is *cellulose thinner*, which has a low flashpoint, and the ease with which it can be ignited is not always appreciated. Obnoxious vapours given off from the solvent as it evaporates produce a light-headed feeling if inhaled by an operator who is not wearing a proper face mask—although this is a short-term effect, lung irritation will be longer lasting. Spray booths should also be well ventilated with an adequate extraction system.

In an enclosed space like a stoving oven, the build up of vapours as the paint dries can cause explosions. To prevent this, an extraction system must be fitted, which will provide a free circulation of air within the oven.

SELECTING THE APPROPRIATE PROTECTIVE TREATMENT FOR SPECIFIC METALS

When considering the appropriate treatment for a specific metal it is wise to consider where the component is to be used, the reasons

| METAL | TREATMENT | | |
	Plated	Hot Dip	Spray or Brush
Mild steel	Chromium Nickel	Galvanise	Stoving Paint
Cast iron	Cadmium prior to zinc		Paint
Aluminium	Anodise (anodic oxidation)		Paint Lacquer
Brass	Chromium		Polish and lacquer

for treatment and *economics* in relation to the application of the component. A very simple form of surface protection is by applying beeswax and polishing with a cloth. This may be completely satisfactory under certain operating conditions but absolutely useless under other conditions.

As with all other treatments, to achieve maximum success it is vital that surface preparation be carried out very carefully.

Electrical *copper* wire can either be solder coated or tinned to prevent any chemical interaction between the rubber or plastics insulation and the copper, and to make soldering easier.

PART TWO

INTRODUCTION TO PROCESS NOTES

The intention of this work is to bring the student into contact with the tools, instruments, materials and machines as soon as possible, so as to make the ensuing lectures more meaningful.

Obviously, a suitable introductory period must be given over to practical demonstrations of the potential dangers and the measures required for safety in the workshop.

The following work programmes are devised to create learning situations and are couched in 'industrial planning sheet' terms for quick assimilation and progress towards the finished work. It would be wrong to expect this practical work to be of a high standard. If the student can see what happens during a series of operations, the subsequent classroom work will have greater significance and more lasting value. The notes will broaden the subject matter and help to provide answers to the questions that are set to be answered after completing the work.

The number of processes prepared provide some selection for preferred choice or to suit the available equipment in the workshop. The last three processes are outside the Unit content but are included for the obviously advanced student who needs additional interest during the workshop sessions.

The prepared processes do not cover the whole of the Unit content but deal with those parts that can best be dealt with in a practical way, leaving other subjects for classroom lectures, providing change and variety in the work programme.

Process 1

Centre lathe, turning between centres. (Figure 1.1)

Object

To observe and note the construction of the tailstock and provisions made for 'across the bed' adjustment, to produce parallel and tapered work between centres.

NOTES ON PROCESS 1: TURNING PARALLEL

While the problem in this process is primarily concerned with turning work parallel, the adjustments necessary to correct tapering are the same, only in the reverse order, for producing taper. Tapers turned by offsetting the tailstock can be as long as the full length of the lathe between centres.

A look at the construction of the tailstock will show that there is a limit to the 'cross-way' movement, so the angle obtainable by this movement is related to the length of the work. Thus on a short shaft a wide angle is obtainable, while on a long shaft the angle is proportionally reduced. Another point that should be noted is the poor initial bearing surface of the work centre holes on the misaligned lathe centres, particularly on short work with a large set-over. A little time is required to allow the centre holes to bed down.

The scale shown on the end face of the tailstock (figure 1.4) provides a guide for zero or for offsetting rather than any specific angle. Several factors, such as wear on the tailstock bed slide-ways, loose-fitting tailstock spindle or even an incorrectly ground centre, can cause the scale to show a false reading. Therefore trial test cuts are necessary before parallel work can be obtained.

When work is turned between centres there are some small but important points to be observed. Among the first to be checked is the concentricity of the live *centre. In theory it should run true, but small burrs in the spindle bore, or on the sleeve and centre, brought about by mishandling, can cause the centre to run out of true. This will show up when the work is reversed on the centres (that is, when the live centre end is placed on the dead centre), to finish turn the end formerly under the carrier. Where the two turned ends meet they will be eccentric. If the instructions in the work sheet are followed this error will be avoided.*

The heat generated by machining causes the shaft to expand so that constant attention has to be given to the free running of the work between the centres, by adjusting the tailstock, and to the lubrication of the dead centre. Work turned between centres is not so rigidly supported as, say, between chuck and centre, so that the rate of stock removal is less and this is further curtailed when the work is slender.

Thin shafts can be supported by a travelling steady but the bearing points cannot be applied until the taper has been eliminated, otherwise there would be a changing reaction from the steady as it moved along the tapering shaft. When a new cut is started at the tailstock end there is enough support from the work centre to allow the cut to travel for a short distance without the steady. This provides a space for the bearing points to be adjusted to the new diameter immediately behind the tool, thus giving support for the work where it is most needed.

PROCESS 1 CENTRE LATHE, TURNING WORK PARALLEL BETWEEN CENTRES

Object   To observe and note the construction of the tailstock and the provision made for 'across the bed' adjustment to produce parallel or tapered work between centres.

Material   Round bar 25 mm diameter × 175 mm long, bright drawn mild steel. The end faces to be machined and centred.

Tools and equipment   Centre lathe, headstock spindle sleeve, 'soft' live centre, dead centre, driving plate, work carrier, right-hand turning tool, 0–25 mm micrometer, plunger-type dial indicator with magnetic stand, 300 mm steel rule.

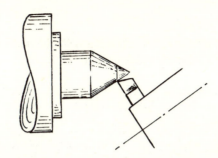

Fig 1.2

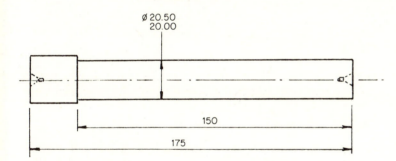

Ø 20.50
20.00

150

175

Fig 1.1

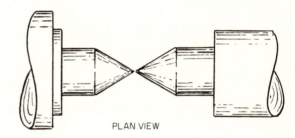

PLAN VIEW

Fig 1.3

Instructions

Machine Setting

Prepare lathe for turning between centres. Fit and check live centre for concentricity; remachine if necessary (figure 1.2). Fit dead centre and 'offer up' to live centre (figure 1.3). Check in plan and elevation. Inspect scale on rear face of tailstock. It should register as shown on the 'inset sketch' of figure 1.4.

Lubricate dead centre hole with grease or tallow, fit carrier to work, and assemble work between centres. Tail of carrier must engage on driving pin of driving plate. Work must rotate freely on

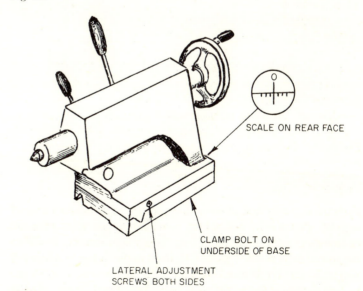

SCALE ON REAR FACE

CLAMP BOLT ON
UNDERSIDE OF BASE

LATERAL ADJUSTMENT
SCREWS BOTH SIDES

Fig 1.4

centres but without any axial movement. Some further adjustment will be required since the work expands by the heat generated in machining.

Machining

Adjust speeds and feeds to produce an acceptable commercial finish. Fit tool and take trial cut 150 mm long. Measure diameters at each end of turned portion. The difference in these measurements shows amount of taper in 150 mm. To correct this error, allowance must be made for unturned portion under carrier, that is, calculate what the difference would be in 175 mm. This is shown and explained in figure 1.5.

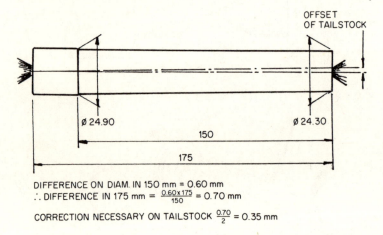

DIFFERENCE ON DIAM. IN 150 mm = 0.60 mm
∴ DIFFERENCE IN 175 mm = $\frac{0.60 \times 175}{150}$ = 0.70 mm

CORRECTION NECESSARY ON TAILSTOCK $\frac{0.70}{2}$ = 0.35 mm

Fig 1.5

Adjustments Necessary to Correct Error

Position dial test indicator so that plunger registers on tailstock spindle, plunger in horizontal position (figure 1.6). Ease off tailstock clamping bolt and adjust 'lateral adjustment screws' on the front and rear face (figure 1.4). In example shown in figure 1.5, dial must register movement away from the operator of 0.35 mm by slackening rear screw and tightening front screw. Secure tailstock clamping bolt, recheck the running clearance of the work

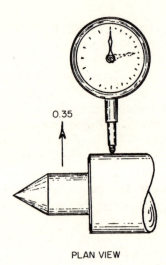

PLAN VIEW

Fig 1.6

on the centres and take another trial cut. Repeat if necessary until parallel work is obtained and then turn to tolerance diameter.

Questions: Process 1

(1) Heat generated by the cutting action of the tool and the friction on the tailstock centre causes the work to expand. What machine adjustments are necessary during machining?

(2) If the live centre does not 'run true' what effect will this have when the work is reversed on the centres (that is, when the unmachined part formerly under the carrier is placed at the tailstock end)?

(3) How can the ends of shafts be faced when the work is supported on the lathe centres?

(4) Why is it not advisable to use a revolving centre for turning between centres?

(5) Make a sketch of a cross-section across the tailstock body to show how the 'across the bed' adjustment works.

Process 2

Centre lathe turning, four-jaw chuck

Object

To produce a concentric round spigot on the end of a square bar. (Figure 2.1)

NOTES ON PROCESS 2: TURNING

If a flat surface is the basic datum face for all precision work in engineering, then the next important geometrical surface must be cylindrical, since it forms such a large part of engineering manufactures.

The work rotates on the same axis as the lathe spindle and should produce true cylindrical form. You will learn later that this is not necessarily so, because fine measuring instruments have been developed that reveal contours that are far from round. Later, during the course, you will have an opportunity to examine samples demonstrating this point. For the moment, it may be assumed that the work produced in a lathe when sliding is, for all practical purposes, cylindrical.

It is impractical to try to set a square bar true in a chuck by testing across the corners of the bar. Figure 2.3 shows rough truing. The corners have uncertain accuracy, produced in the course of manufacture or in stock handling or in setting. They are in theory the junction of two planes at right-angles and constitute a line without dimension except for length.

Self-centring four-jaw chucks are available but they are rarely part of the standard equipment of a plain centre lathe. These are to be found in the production workshops where large numbers of identical workpieces are made at a minimum cost. A device for holding the square bar securely and true, called a fixture, could be made, but the cost of such a fixture would not be justified for one-off or even a small job lot. This is clearly a case for employing the skill of a trained operator.

Turning across the corners of a square bar has the effect of a broken cut, which causes the chips to spray out further than they would do with a continuous cut. Operators engaged on this work should take additional safety precautions by erecting a screen to deflect the chips, as well as wearing goggles.

The vibration set up by a broken cut could also upset the initial concentric setting of the work so that the depth and feed of cut have to be regulated to reduce this possibility to a minimum. Equally important, controllable features such as work and tool overhang should be as small as possible when setting up for machining. Non-working slides might also be tightened where suitable clamping screws are provided.

An example of this type of work might be found on a trunnion block of a heavy lorry transmission shaft or on the jib pivot of a crane where the alignment of two opposite spigots is important.

PROCESS 2 CENTRE LATHE TURNING, FOUR-JAW CHUCK

Object   To machine a concentric spigot on the end of a square bar (figure 2.1).

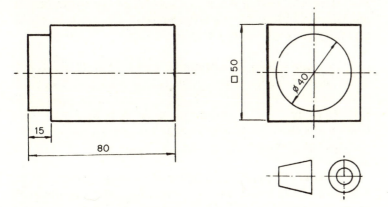

Fig 2.1

Material 50 × 50 mm square bar 80 mm long one-off, bright drawn mild steel. Suitable packing strips for protection of work.
Tools and equipment Centre lathe, four-jaw chuck, scribing block, spirit level, plunger dial gauge with tool post mounting, 150 mm steel rule, 25–50 mm micrometer.

Instructions

Prepare the centre lathe and four-jaw chuck for assembly, making sure that all mating and locating faces are clean and free from burrs and foreign bodies. Chuck jaws should be fitted as shown in figure 2.2.

Machine Setting

Grip work over protective packing, leaving 40 mm of material protruding from chuck jaws. Rough true by using concentric rings

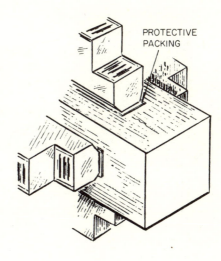

Fig 2.2

on chuck face. Using the scribing block in a convenient position on the lathe bed, adjust the scriber point to centre height.

Set material approximately true by rotating chuck and testing across corners of work with scriber point (figure 2.3). Final truing is achieved by using spirit level and dial test indicator (d.t.i.). Clamp d.t.i. horizontally in tool post with plunger at centre height. Place spirit level on work as in figure 2.4 and rotate chuck until level reads zero. Feed in cross-slide to enable d.t.i. plunger to contact face A, figure 2.4. Set d.t.i. to a convenient zero. Record cross-slide indexing position. Withdraw cross-slide so that d.t.i. clears work. Rotate work 180°. Replace spirit level as in figure 2.4 and rotate chuck until level reads zero. Feed in cross-slide to same position as for face A. Note d.t.i. reading. Adjust chuck jaws so that readings for faces A and C are the same. Repeat for faces B and D.

Machining

Because of broken cut use safety glasses. Adjust lathe to run at a suitable speed and feed. Face off work. Continue to turn a spigot

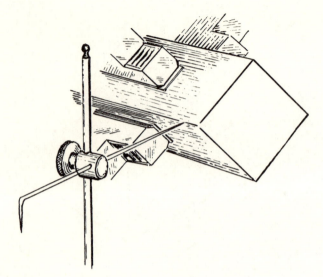

Fig 2.3

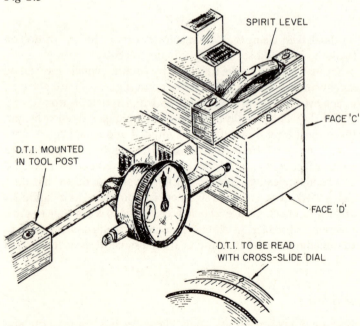

SPIRIT LEVEL

FACE 'C'

FACE 'D'

D.T.I. MOUNTED
IN TOOL POST

D.T.I. TO BE READ
WITH CROSS-SLIDE DIAL

Fig 2.4

approximately 15 mm long and 40 mm diameter. Produce a good commercial finish, remove all burrs and sharp edges.

Check for concentricity.

Questions: Process 2

(1)　State why testing across the corners of the square bar is unlikely to produce the required degree of concentricity of the turned spigot to the faces of the square bar.

(2)　How can a dial test indicator be used to indicate the correct adjustment of the chuck jaws when setting work?

(3)　When machining across the corners of the square bar, causing a broken cut, what particular hazard does this make for the machine operator?

(4)　A broken cut tends to set up vibration in the machine. What factors in tool and work setting would reduce this to a minimum?

(5)　By means of a simple sketch show how the spigot was tested for concentricity after the work had been removed from the machine.

Process 3

Taper turning (compound slide)

Object

To machine matching external and internal tapers. (Figure 3.1)

NOTES ON PROCESS 3: MATCHING TAPERS

Tapers produced by the method described in this process are subject to minor errors, owing to the difficulty of maintaining a uniform feed in manual operation. The quality of the finish and accuracy are further impaired by excessive clearance in the slide when changes in hand pressures on the operating handle cause fluctuating pressures on the tool, particularly on light finishing cuts. Nevertheless, good-fitting tapers are obtainable when the conditions are right.

In general the angular setting scales on the compound slide base are seldom finer than half a degree and are not comparable with the accurate settings obtainable on the taper-turning attachment scales.

When large angles are involved the height of the tool above or below the work centre can affect the angle produced, so care should be exercised to see that the point of the tool is correctly set to the work centre.

It is difficult to measure a taper in the machine, but reasonable results can be obtained using a knife vernier over an established length, bearing in mind that the length measurement between the two diameters is critical. Better results might be obtained by gauging the 'rise' of the taper on a dial indicator over a slide indexed length parallel to the work axis. Final measurement should be made on a surface table with standard rollers, slip gauges and a micrometer. When a test gauge is available, satisfactory checks can be made while the work is still on the machine using engineers' blue to obtain a marking.

The over-all length of a taper produced by this method should not be greater than the length of the slide movement. Resetting the slide to turn a longer taper is not satisfactory unless the work is to be finally ground.

The use of matching tapers for the temporary or permanent assembly of tool components has a widespread use in machine tool engineering. A very good example of the application of both types of taper may be seen in the design of a popular type of drill chuck. The body of the chuck is secured to the shank 'permanently' by a short length self-holding taper while the shank is held in the tapered bore of the drilling machine spindle 'temporarily' by another self-holding taper.

Frictional resistance causes these tapers to hold together and the

angles to which they are ground are precisely controlled in relation to the diameter of the spindle. Standard tables have been compiled for the selection of these factors, notably Morse *and* Brown and Sharpe *tapers.*

PROCESS 3 TAPER TURNING (COMPOUND SLIDE)

Object   To machine matching external and internal tapers (figure 3.1).

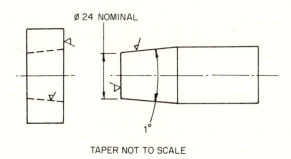

Fig 3.1

Material   Bright round mild steel bar 50 mm diameter × 20 mm long one-off. Bright round mild steel bar 30mm diameter × 75 mm long one-off.

Tools and equipment   Centre lathe, boring bar, right-hand turning tool, plunger-type dial test indicator, engineers' marking blue, drill chuck, centre drill, 23 mm drill.

Instructions

External Taper

Set 30 mm bar in three-jaw self-centring chuck so as to leave 45 mm protruding from chuck jaws.

Fit turning tool on centre height. Face off work. Unclamp compound slide and index $\frac{1}{2}°$ in an anti-clockwise direction. Reclamp. Rough machine a taper until an end diameter of 25 mm is reached. Finish turn to a diameter of 24 mm (figure 3.1).

To obtain good location from mating tapers it is important to produce a high standard of surface finish from the tool.

Do not remove from chuck

Setting for Internal Taper

Remove turning tool and fit boring bar.

Attach dial test indicator to shank of boring bar as in figure 3.2. Unclamp compound slide and index 1° in clockwise direction. Reclamp. Cross-slide can now be adjusted so that plunger of dial test indicator locates on work as in figure 3.2.

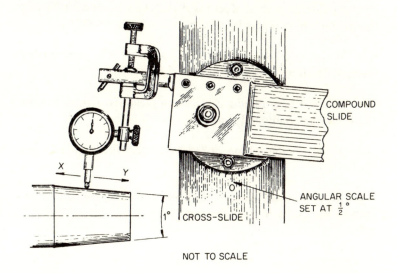

Fig 3.2

Traverse compound slide along line XY (figure 3.2) and observe readings on dial test indicator. Adjust angular position of compound slide until X and Y read zero.

Machining Internal Taper

Remove external taper and dial test indicator. Place 50 mm diameter material in chuck. Face, drill and bore hole 23 mm diameter through work. Machine internal taper taking roughing cuts until bore is tapered along its length.

Offer external taper into bore. Continue to machine until external taper just enters.

Complete machining operation with finishing cuts. Good surface finish is important. Remove sharp edges.

Lightly smear engineers' blue over external taper and take a light rubbing in bore to test contact area.

Questions: Process 3

(1)   Could this method of setting the machine be used for turning longer matching tapers by a taper-turning attachment?

(2)   What steps should be taken to ensure the good-quality surface finish necessary for matching tapers?

(3)   What degree of angular accuracy can be obtained from indexing the compound slide?

(4)   How can the setting of the tool affect the angle of taper produced?

(5)   Show how an accurate diameter measurement can be obtained on, say, a Morse taper sleeve.

Process 4

Screw cutting (single-point tool)

Object

To produce a machined screw thread (figure 4.1) by setting compound slide half thread flank angle; observing important features of the screw thread, and noting the principles of forming and generating.

NOTES ON PROCESS 4: SCREW CUTTING

When cutting screw threads on a centre lathe with a single-point tool the thread is apt to tear owing to the swarf jamming on the tool point. If the tool is fed in square to the work axis, the swarf flows from both flanks of the thread and so tends to obstruct the flow from each face.

An improved method of in-feeding can be obtained by setting over the compound slide to half the thread flank angle, adjusting the toolpost so that the tool is positioned square to the work axis, as before, and feeding the cuts on the angled compound slide. In this way the bulk of the cut is taken on one flank of the thread, allowing a free flow of swarf, while the opposite face is lightly scraped by the form of the tool.

The next problem is in setting the lathe to cut the required pitch of screw thread. Consider the lathe in its simplest condition. A train of equal-size gearwheels is arranged to couple the drive between lathe spindle and leadscrew. If the pitch of the leadscrew is, say, 6 mm, then the pitch of thread produced on the work will also be 6 mm when the tool and the leadscrew nut are engaged. By setting different size gearwheels to alter the ratio between work spindle and leadscrew revolutions, a variety of pitches is obtainable on the work. For example, a ratio of 2 to 1 (the smaller gear on the work spindle) would produce a pitch of 3 mm on the work. Fortunately this problem does not arise on modern machines, since a gearbox is interposed between the lathe spindle and the leadscrew, which provides for a selection of pitches by the simple operation of gearbox selectors, but the principle should be understood.

There remains the problem of 'picking up' the thread at the beginning of each cut. Where the pitch of the thread to be cut divides evenly into the pitch of the leadscrew, the tool will always start at the right point. Take, for example, a $1\frac{1}{2}$ mm pitch thread to be made on a lathe having a leadscrew of 6 mm pitch. Every pitch of the leadscrew 'lines up' with every fourth pitch of the work screw. This allows the leadscrew nut to be engaged in any position. In the case of a $1\frac{1}{4}$ mm pitch thread to be made on the same lathe, the work thread would only 'line up' at every fifth pitch of the leadscrew. In other words, 24 pitches of the work evenly match 5 pitches of the leadscrew so that a point has to be selected to equal these conditions before starting each cut. Further examples are shown in the diagrams for this process.

A chasing dial fitted to the saddle of the lathe solves the problem of selecting the right point to engage the leadscrew nut. This attachment consists of a circular dial indexed round the circumference. The dial is secured to a spindle, which passes through the attachment and is free to rotate. At the lower end of the spindle provision is made for fastening a small gearwheel having the same tooth pitch as the leadscrew. The gearwheel meshes with the leadscrew, which causes the dial to rotate when the leadscrew is in motion and the saddle is at rest. Lathe manufacturers usually provide a chart attached to the apron of the machine to show the size of the gearwheel to fix on the chasing dial spindle and a mark on the dial to engage the leadscrew nut for any particular pitch of screw thread. The principle, however, should be understood in that leadscrew rotation is shown by the chasing dial in relation to work rotation.

A screw thread made by a single-point tool will not be the correct form until the crest has been rounded off by a hand or machine chaser of the same form. The point of the tool has to be oilstoned to give the required radius at the root of the thread. A screw pitch gauge is the best guide for this operation while a thread tool gauge (illustrated in the diagrams for the operation schedule) is used for obtaining the flank angles when grinding the tool and setting the tool to the work.

A recess at the end of a thread is the most convenient arrangement for the tool run out. The width should be about $1\frac{1}{2}$ times the thread pitch and the same depth as the thread. Another method would be by withdrawing the tool at the required length while the lathe is in motion. A succession of cuts leave a tapering effect but this requires some practice to obtain a smooth finish.

There are several types of machine chaser, which make better tools for cutting threads, but the cost has to be considered since each thread form requires a separate chaser. A single-point tool will produce any form of thread by a simple grinding operation at the tool point. Machine dies are also used although it is not always convenient to set them up on a centre lathe.

PROCESS 4 SCREW CUTTING (SINGLE-POINT TOOL)

Object   To demonstrate the principles of forming and generating in the machining of a screw thread.

Material   Bright drawn mild steel bar 25 mm diameter × 100 mm long.

Tools and equipment   Centre lathe suitable for screw cutting, right-hand turning tool, recessing tool 3 mm wide ($1\frac{1}{2}$ × pitch) 60° vee-form screw cutting tool, thread tool setting gauge, 0–25 mm external micrometer, 150 mm steel rule.

Instructions

Grip work in three-jaw self-centring chuck; allow bar to project 50 mm from chuck jaws. Face off. Turn 24 mm diameter for a length of 31 mm. Turn boss and recess to position and diameter shown in figure 4.1.

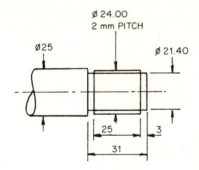

Fig 4.1

Machine Setting

Unclamp compound slide (top slide) and rotate 60° (see figure 4.2). Clamp and recheck angular setting. Set single-point screw cutting tool in toolpost on centre height. Swing toolpost round to bring 60° vee-form square to work axis, check with vee-form setting gauge (figure 4.3). Clamp toolpost and recheck tool position.

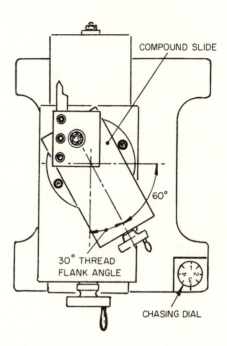

Fig 4.2

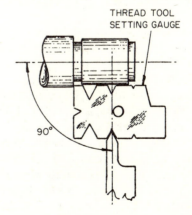

Fig 4.3

Preparation for 'In Feed' of Tool

Lathe gearbox ratio should now be adjusted to produce a pitch of 2 mm. With lathe spindle running at a suitable r.p.m. 'kiss' work with tool. Set both cross-slide and compound slides indexing dials to zero, ensuring all backlash is removed.

Preparation for Engaging Leadscrew Nut

If pitch to be cut is a multiple of leadscrew pitch the nut may be 'dropped in' at any point. If pitch to be cut is *odd*, that is, it will not divide evenly into the leadscrew pitch, read the instructions given in the machine handbook or on a plate usually provided on the machine apron.

Reversing the lathe spindle while the leadscrew nut remains engaged is an easy alternative but defeats the object of understanding the principles and use of the chaser dial mechanism.

Machining

Run the lathe spindle at a moderately slow speed to stop the thread tearing. Position tool clear of end of work with cross- and compound slides set at zero, as explained earlier. Engage leadscrew nut and by a series of light cuts on the compound slide form and generate thread until point of tool touches minor diameter boss 21·40 mm.

Allow tool to run into recess at end of each cut and then quickly withdraw tool and disengage leadscrew nut at same time. Some practice at this before actual thread cutting may reduce tool resetting by fouling the shoulder of the recess. Mean depth of thread may be calculated at 0.70 × pitch subject to radius on point of tool being 0.06 × pitch. This is best gauged on a *thread pitch gauge* or the same pitch chaser. The indexed compound slide movement will be slightly more as it moves down the 30° flank.

'Even and odd' threads (figures 4.4a and b)—for the 'even' threads it will be seen that any beginning point of the leadscrew will line up with thread being cut. 'Odd' threads require three pitches of the leadscrew before lining up with thread being cut.

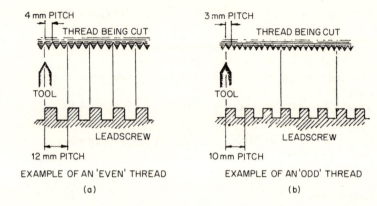

Fig 4.4

Questions: Process 4

(1)   The method of machining the thread might be said to form one flank and generate the other. Explain this.

(2)   Referring to the position of the compound slide, describe another method of single-point screw cutting.

(3)   Compare the advantages and disadvantages of the methods in your answers to questions 1 and 2.

(4)   The termination of the thread in a recess is an accepted method. Name two other methods of finishing a thread.

(5)   Name two methods of producing screw threads on a centre lathe other than by using single-point tools.

Process 5

Turning eccentrics. (Figure 5.1)

Object

To study methods of offsetting work to produce eccentrics, either to open tolerances or close limits.

NOTES ON PROCESS 5: TURNING ECCENTRICS

An eccentric shaft has a centre boss or end bosses, which are offset from the axis on which it rotates. The item to be made (figure 5.1) rotates on its 50 mm diameter axis. The offset ends could then be made to impart a reciprocating motion to connecting-rods suitably bored and supported at right-angles to the shaft axis.

The eccentric ends are 'timed' to operate at 180° to each other, but any other arrangement of timing could be made. In this context, the marking-out plays an important part in the location of the eccentrics. Where great accuracy is not required the work could be set and finally machined to the marking, anticipating a tolerance of ±0.14 mm.

For finer tolerances toolmakers' buttons would be used by drilling and tapping the centre of the eccentrically marked faces to take the button clamping screw. The buttons would then be lightly fastened on the work faces and the final setting carried out on the surface table using a vee-block, slip gauges, dial indicator and the necessary calculations. The work can then be set up in the lathe and the eccentrics set true by means of the buttons.

The alignment of the axes parallel to each other requires care especially when the eccentrics are spaced further apart axially. This setting operation is greatly assisted if the chuck is in good condition. The jaw slides must not have excessive wear, since this allows the jaws to hinge backwards so that they cannot hold the work uniformly along the full length of the gripping face. Damaged or distorted jaws will also add to the difficulty of setting the work.

For a given set-over of an eccentric, the dial gauge reading in the lathe will show double the amount of set-over. This makes the reading very critical and also curtails the scope of the dial gauge because of the limited plunger movement. Provided that the cross-slide leadscrew and nut are not badly worn, the required dial gauge movement can be transferred to the cross-slide indexing dial so that the dial gauge only needs two zero settings. This application forms part of the instructions in setting the 28 mm diameter eccentric on the work sheet.

Where the total set-over exceeds half the width of the chuck jaw, some sort of holding device has to be made for the two opposite jaws at right-angles to the common centre-line of the work. Such a device is shown in figure 5.3. This allows uniform pressure from each jaw to act on the work and also greatly assists work setting. Where a quantity of similar eccentrics has to be made, the cost of a properly designed fixture would be justified because of the time saved in setting and for the maintenance of consistent accuracy in the relation of the eccentrics to each other.

PROCESS 5 TURNING ECCENTRICS

Object   To demonstrate methods of offsetting work to produce eccentrics either to open tolerances or close limits.

Material   Bright mild steel round bar, 50 mm diameter × 90 mm long, faced at both ends and marked out to dimensions given in figure 5.1. (See process 11.)

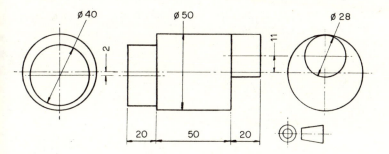

Fig 5.1

Tools and equipment   Centre lathe, four-jaw chuck, protective packing for work, surface gauge, 25–50 mm micrometer and 150 mm steel rule, dial test indicator (5.0 mm range) with toolpost mounting, special split block (figure 5.3) for holding stock while machining 28 mm diameter eccentric, right-hand turning tool.

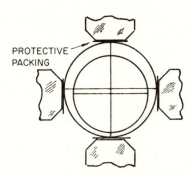

Fig 5.2

Instructions

Work Setting

40 mm diameter eccentric. Grip stock over protective packing, allowing 30 mm to project from jaws. Set lines on work to jaws of chuck (figure 5.2). This facilitates work setting.

Set work to marking-out lines with the surface gauge. Finally true with d.t.i. obtaining a 4.00 mm 'run-out' over the stock bar. Finish turn 40 mm diameter eccentric.

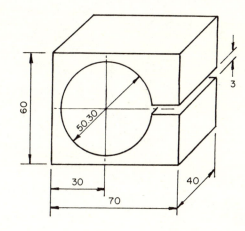

Fig 5.3

Resetting Work

28 mm diameter eccentric. Place work in split block (figure 5.3) allowing 30 mm to project. Line up work centre-lines to chuck jaws and true work to marking-out with surface gauge. Finally, fit d.t.i. to toolpost, plunger horizontal and true by taking a zero reading at maximum throw and set cross-slide indexing dial at zero (figure 5.4). Turn chuck 180°, index cross-slide 22 mm, and take a *minimum* reading of d.t.i. (figure 5.5). If work is correctly set, this should also read zero.

Note   It is highly improbable that a correct setting will be obtained at the first attempt. Repeat the procedure, correcting by

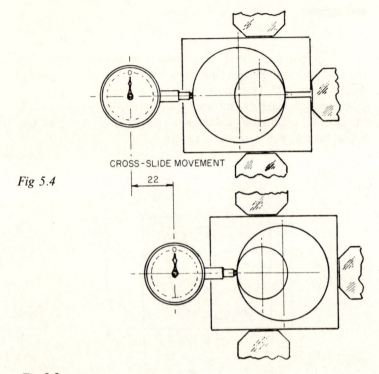

CROSS-SLIDE MOVEMENT

22

Fig 5.4

Fig 5.5

(3)   Suggest another way by which the work may be held for machining the 28 mm diameter eccentric instead of using the split block shown in figure 5.3.

(4)   What would be the probable cause of vibration in the lathe when running at normal cutting speeds, especially when turning the 28 mm diameter eccentric?

(5)   How might toolmakers' buttons have simplified work setting and eccentric alignment?

small adjustments of the chuck jaw until a double zero setting is obtained in conjunction with the cross-slide movement of 22 mm.
    Finish machine 28 mm diameter eccentric taking precautions necessary for a 'broken cut'.

Questions: Process 5

(1)   If the chuck jaws are badly worn in their slides or tapered on their gripping faces (bell mouthed) what effect might this have on work setting?

(2)   Why was the split block used to hold the work for one eccentric but not for the other?

Process 6

Centre lathe, taper-turning attachment

Object

To machine a number 3 Morse taper (figure 6.1) and to observe the advantages and disadvantages of the method.

NOTES ON PROCESS 6: TAPER TURNING

There are several types of taper-turning attachment. In principle they all function on similar lines. The most frequently seen example does not require any major alteration in the lathe design except for additional pads drilled, tapped and dowelled to take the attachments. The really important difference is in the location of the cross-slide thrust bearing, which is of course a two-way thrust.

Where provision is made for an attachment, the thrust bearing is situated on top of the angle setting slide, *(figure 6.2) instead of the usual position hidden in the saddle casting behind the cross-slide indexing dial.*

To make a close study of the mechanism it is advisable to switch the power off and operate the saddle manually.

The angle setting slide should then be set over to one of its extreme positions, that is, at its maximum angle to the lathe bed and secured by the two clamping screws. The attachment lock *may then be fastened to the lathe bed. When the saddle is moved along the lathe bed, the thrust from the angle setting slide can be seen operating laterally through the cross-slide leadscrew causing the slide to move towards or away from the front of the machine, depending on the way the slide is set. Notice how the* support slide *moves along with the saddle immediately below the leadscrew thrust bearing, thus giving support at the point of maximum pressure.*

It should follow from the above that the operating handle of the cross-slide would move with the slide but this is not the case, since an extended boss of the handle is internally splined to take the end of the leadscrew, which is similarly machined. The handle is held against the saddle face by a suitable bearing allowing it to rotate freely and to operate the cross-slide independently. There is sufficient clearance inside the handle boss to allow for end float of the leadscrew.

The only disadvantage in the design of this mechanism is the delay in action at the beginning of each cut, called 'backlash' in the instructions of the process sheet. This is caused by the several clearances that have to be taken up before the tapering movement begins to act. The clearances exist between the sliding faces of the support slide and the angle setting slide. There is also some clearance in the thrust bearing, but probably the greatest cause of delay is in the cross-slide leadscrew nut. When the tool is fed forward for a cut

the pressure from the screw is applied to one face of the thread. Before any action can take place from the angle setting slide the pressure from the screw has to act on the opposite face of the thread, thereby closing up the total clearance between nut and screw. This is the point of greatest wear whether the lathe is used for parallel or taper turning.

PROCESS 6 CENTRE LATHE, TAPER-TURNING ATTACHMENT

Object　To machine a number 3 Morse taper (figure 6.1) and to observe the advantages and disadvantages of the method.

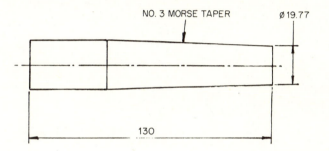

NO. 3 MORSE TAPER

Ø 19.77

130

Fig 6.1

Material　Bright mild steel bar 25 mm diameter × 200 mm long.
Tools and equipment　Centre lathe fitted with taper-turning attachment, standard Morse taper sleeve, engineers' marking blue, 150 mm steel rule, 0–25 mm micrometer, right-hand facing and turning tool, BS 1660: 1972 Machine tapers, reduction sleeves and extension sockets.

Instructions

Preparation

The lathe must be checked for the alignment of the tailstock centre to the headstock. (Refer to process 1, figure 1.3.)

Set work in three-jaw chuck with 10 mm protruding from jaws. Fit facing tool and lightly face end of bar and centre drill. Support end of bar on tailstock centre having previously lubricated work centre hole. This centre will require adjustment and relubricating from time to time, owing to work expansion and centre hole wear.

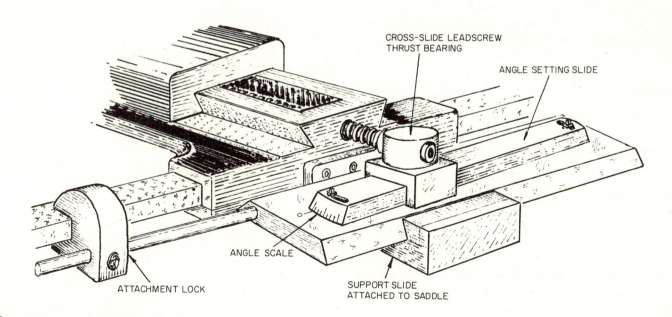

CROSS-SLIDE LEADSCREW
THRUST BEARING

ANGLE SETTING SLIDE

ANGLE SCALE

ATTACHMENT LOCK

SUPPORT SLIDE
ATTACHED TO SADDLE

Fig 6.2

Fit right-hand turning tool taking care to set to correct centre height and move saddle to position tool at right-hand end of work.

Taper-turning Attachment

At this stage attachment lock and shaft should not be fitted. Push angle setting and support slides towards headstock as shown in figure 6.2. This will ensure that there is sufficient 'run' to complete full taper length. Put angle setting slide over to half included angle of work taper and secure. Finally, fit attachment lock and shaft and clamp to lathe bed.

Machining

Because of backlash in the mechanism that operates the taper-turning attachment it is necessary to set cut about 20 mm to right of end of work, to provide a 'run in' to take up backlash. Amount of 'run in' depends on condition of machine.

Take several trial roughing and finishing cuts. Test taper using engineers' blue and standard Morse taper sleeve. Adjust taper setting slide as necessary until a good fit is obtained. Because these tapers are self-holding, a good finish on the final cut is important.

Questions: Process 6

(1) The principle of most taper-turning attachments is that the cross-slide is acted on by another slide nearly at right-angles to it. What component of the cross-slide, shown in figure 6.2, transmits this motion?

(2) In any assembly of slides and moving parts, clearance between the components is necessary. What effect would an accumulation of these clearances have on the immediate actuation of the cross-slide?

(3) If the attachment lock and its connecting shaft were removed, would the saddle and the cross-slide be free to operate in the normal way?

(4) State two advantages that the taper-turning attachment has over the compound slide.

(5) Briefly explain another method of turning long shallow tapers and discuss its advantages and disadvantages relative to the taper-turning attachment.

Process 7

Centre lathe, face plate

Object

Boring an accurately positioned hole in a bearing block, set on an angle bracket mounted on a face plate.

NOTES ON PROCESS 7: FACE PLATE

An important factor in this type of work is in making sure that the components of the set-up are properly balanced and secured. Unbalanced revolving masses present a very serious danger to the operator and, indeed, to all the personnel in the workshop. The work should be balanced when the gears in the headstock are in a neutral position so that the spindle is free to rotate without the drag of the meshing gears. Even when static balance has been obtained, spindle speeds should be kept as low as possible, consistent with satisfactory cutting conditions, since small unobservable errors in the static balance test become a major condition of imbalance at higher spindle revolutions.

The several sharp and irregular projections of the work present additional hazards for picking up loose clothing or in striking part of the saddle or toolpost that might detach or loosen some part of the work set-up.

An alternative set-up for this work in a four-jaw chuck might be thought to be less cumbersome and saving in time, but there are several aspects that need to be considered. The height of the bore centre (dimension B) and the alignment of the bore parallel to the bearing block base are two very important features, which are preset in the face and angle plate arrangement. These are not so easily checked with the work held in a chuck, and if there should be a number of bearing blocks to be machined, the advantages of the preferred method are clear. Chuck jaw pressures can affect the roundness of the finished bore when the work is released, and in the case of a split bearing the pressure from the two jaws in line with the half faces may cause the halves to be misplaced where alignment dowels or stepped faces are not used.

It is unlikely that the bored hole in the bearing block would be the bearing surface in which a rotating shaft operates unless the shaft rotates at very low speeds, since the frictional resistance of steel on steel is comparatively high. It is more likely to house a ball- or roller-race and the fit of the bearing outer case in its housing plays an important part in the running efficiency of the race. Makers of ball- and roller-races specify close tolerances for the housing bore and these should be adhered to, since they are designed to take account of the compressive forces and the heat generated in the race under running conditions.

Where plain bearings, such as phosphor bronze liners are fitted, reference should be made to BS 4500: Selected I.S.O. fits in the transitional or interference fit range.

PROCESS 7 CENTRE LATHE, FACE PLATE

Object   Boring an accurately positioned hole in a bearing block, set on an angle plate mounted on a face plate.

Material   Bright mild steel bar $50 \times 100 \times 125$ mm milled to dimension given in figure 7.1 and rough bored to 35 mm diameter.

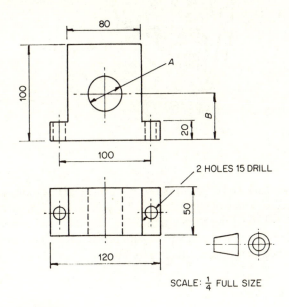

SCALE: $\frac{1}{4}$ FULL SIZE

Fig 7.1

Tools and equipment   Centre lathe, face plate, angle plate, counter-balance weights, slip gauges, alignment test mandrel to suit lathe spindle, dial test indicator, surface plate, telescopic gauges, spirit level, boring bar, test plug.

Instructions

Mount face plate on lathe spindle.

Clean internal taper to lathe spindle and insert mandrel (figure 7.2). Attach angle plate on face plate to approximate position

required by dimension *B* on bearing block. (For dimension *B* see note at foot of this page and figure 7.1.)

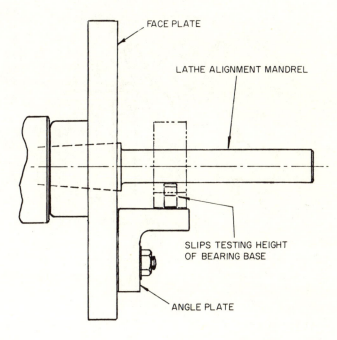

Fig 7.2

Machine Setting

Wring slip gauges together to the calculated value.

$$\text{Slip gauge value} = \text{distance } B - \frac{\text{diameter of mandrel}}{2}$$

Use slip gauges to set angle plate relative to mandrel. Secure to face plate and recheck.

Attach work as in figure 7.3 and adjust so that vertical centre-line of bore lies on centre axis of lathe in the 'across the bed' direction. To do this refer to process 2, figure 2.4, for setting square workpiece in four-jaw chuck with spirit level and dial test indicator (figure 7.3).

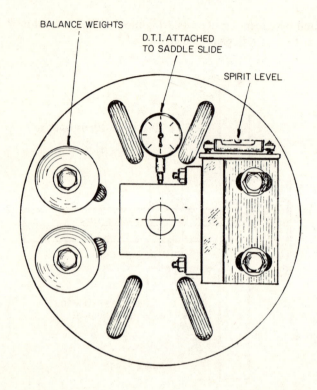

Fig 7.3

Secure dial test indicator to toolpost and set plunger to locate on front face of work. Check alignment of front face of bearing block parallel to cross-slide motion (figure 7.4).
Secure work.
Attach counter-balance weights to face plate.

Machining

Set boring bar in toolpost. Rough machine bore.
Finish machine bore to dimension A using telescopic gauges and external micrometer to take measurements.

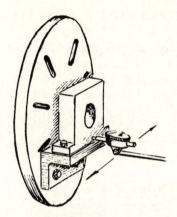

TRUEING FACE OF BEARING BLOCK

Fig 7.4

Checking

Remove work and check dimension B using dial test indicator and slip gauges. Value of slip gauges $= B - \frac{1}{2}$ diameter of finished bore. *Note* The bearing block (figure 7.1) can be used continually provided that a new diameter A and centre height position B are established before the student sets up.

Questions: Process 7

(1) Explain why both angle plate and face plate are made of cast iron.

(2) Why is it necessary to attach counter-balance weights?

(3) Could the bearing block have been machined by holding it in a four-jaw chuck? If so, how could the centre of bore A in figure 7.1 be located?

(4) How could the position of the bore be checked without a test plug?

(5) State four possible sources of danger in this type of work.

Process 8

Shaping a vee-groove

Object

To observe (1) the relative positions of the machine front slide and clapper box, and (2) the method used to obtain accurate depth and position.

NOTES ON PROCESS 8: SHAPING A VEE-GROOVE

In a general engineering workshop the shaping machine is a useful and versatile tool that is not always given the maintenance it deserves. Compared with a milling machine it is an inexpensive tool and its running requirements, in the way of cutting tools and equipment, make it very economical.

In a small way, the work in this exercise demonstrates a degree of versatility when three faces at different angles to each other are machined in one work setting and with the same single-point tool.

An important feature is the function of the clapper box, in the way it allows the tool to lift by the hinge of the clapper on the backward stroke of the ram, thus saving wear on the clearance face of the tool. Some machines are now fitted with an automatic lifting device, which further improves the life of the tool.

Some study should be given to the reason for changing the relative position of the clapper box to the front slide when machining angled faces. This is best observed while the cutting is in progress. The axis of the clapper hinge has to be less than a right-angle, seen from the lower end of the clapper, to allow the tool to swing clear in an arc from the face of the work. To put it more simply (or to establish a rule): the point of the tool should be swung towards the face being machined. The instructions on the work sheet show how this is done and the principle also applies when machining vertical faces. Wear on this important part of the machine occurs at two places: (1) on the clapper hinge pin and (2) at the clapper housing faces. Where they exist they seriously impair the quality of the finish.

The cross-slide feed mechanism is easier to follow but there are some features that are not immediately obvious. The timing of the interrupted feed operation should not take place during the forward cutting stroke, which is also connected to the amount of feed since adjustments are made at the same point. The amount of feed per stroke is regulated by the radius of the feed crank pin mounted on the tee-slotted face of the feed crank wheel. If the crank pin is moved outwards the swing of the ratchet quadrant on the end of the cross-slide leadscrew is increased thus engaging more teeth of the ratchet wheel by the spring-loaded plunger.

The arrangements made for maintaining a uniform distance between the cross-slide leadscrew and the centre of the feed crank wheel at any position of the vertical slide should be noted. Contained in a pivoted housing are two gears—one on a fixed bearing transferring power from the inside of the shaper casing to another gear, which has an extended shaft through the housing to take the feed crank wheel. The assembly then is able to swing on its 'fixed' bearing to take up any difference between the centres spanned by the connecting-rod.

PROCESS 8 SHAPING A VEE-GROOVE

Object   To observe (1) the relative positions of the front slide and clapper box, (2) the method of obtaining accurate depth and position measurements.

Material   Bright mild steel rectangular bar 30 × 50 × 75 mm long, the end faces (30 × 50 mm) to be machined square and a 3 mm slot milled through the centre as shown in figure 8.1.

Tools and equipment   Shaping machine, machine vice, parallel packing pieces as necessary, shaping tool as shown in figure 8.2, depth micrometer, set of metric slip gauges, Hoffman roller 20 mm diameter, surface plate, vernier height gauge, 150 mm steel rule.

Instructions

Calculate distance across top face of vee [2(tan 43° × 12)] and mark lines evenly spaced about centre slot as shown in figure 8.3.

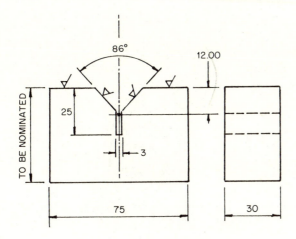

Fig 8.1

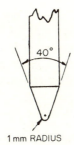

Fig 8.2

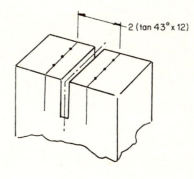

Fig 8.3

Machine Setting

Fit vice to machine work table with jaws square to ram slide. Place work in vice on parallel strips allowing 18 mm to project above vice jaws and secure. Set tool square in toolpost. Adjust stroke to 40 mm (see scale usually placed on upper surface of ram slide).

Position stroke to give overlap of 5 mm from each work face (figure 8.4).

Machining

Take a light finishing cut across top face. (It is anticipated that the dots of the marking will show through this cleaning-up cut.)

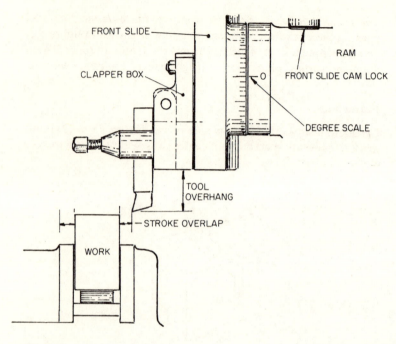

Fig 8.4

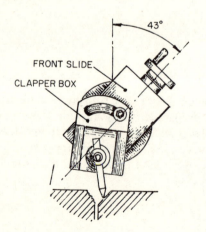

Fig 8.5

Machining Right-hand Angle Face

Swing front slide 43° to right and secure. Adjust clapper box as shown in figure 8.5. Reset tool to give necessary clearance and rough machine face to marking.

Machining Left-hand Angle Face

Swing front slide through 86° to left. Reset clapper box and tool as shown in figure 8.6, and rough machine left-hand face.

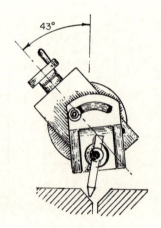

Fig 8.6

Measurement

Run work clear of tool and slide, clean top vee-faces, assemble roller, slips and depth micrometer (figure 8.7) to measure depth of roller in vee. The calculations involved are shown in figure 8.8. Finish machine left-hand face to marking.

Finishing to Size

Swing front slide, clapper box and tool to complete right-hand face. Take a series of light cuts on face, assembling measuring instruments after each cut to check depth. Note that the roller is

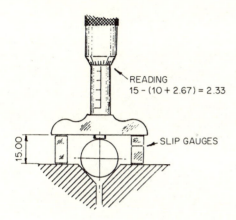

Fig 8.7

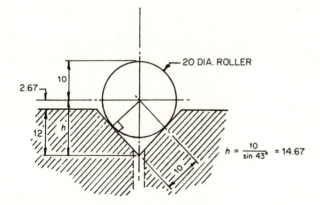

$$h = \frac{10}{\sin 43°} = 14.67$$

Fig 8.8

lowered approximately one-and-a-half times the amount taken off the angle face. (Refer to figure 8.8).

Questions: Process 8

(1)   What combined slide movements produce: (a) the top flat surface (b) the vee-faces?

(2)   If the clapper box were not positioned as shown in figures 8.4 and 8.5 when machining the angled faces, what effect would this have on (a) the work finish (b) the cutting action of the tool?

(3)   Explain why the method shown for measuring the vee-depth (figures 8.6 and 8.7) is used.

(4)   (a) Why is the backward movement of the ram made faster than the forward movement? (b) Excluding hydraulics, name a mechanism used to obtain this difference in speed.

(5)   Explain how the vee could be machined on a horizontal milling using standard cutters.

Process 9

Shaping

Object

To examine the problems of machining dovetail slides and the function of the clapper box, particularly on the 60° inverse face. Also to test a method of accurate measurement of the slide contact faces.

NOTES ON PROCESS 9: SHAPING A DOVETAIL

The problems of machining the short half dovetail in this process are the same as, say, machining a lathe cross-slide. The length of the slide to be machined should not exceed three-quarters of the maximum stroke, since accuracy tends to fall off at the ends of the stroke. Work beyond the capacity of the shaper should be transferred to the planing machine.

The main difficulty is in machining the 60° inverse face, since manual feeding is tedious and does not always produce a good finish. Where the front slide is equipped for power feeding the problem does not arise, but the correct setting of the clapper box is critical and unless it is properly set, as in the instructions on the process sheet, machining will be impossible.

The device used for saving time on the non-cutting return stroke of the ram should be studied, by first switching off the power to the machine and then removing the inspection cover to expose the slotted link mechanism. *A long elliptic-shaped casting, called the slotted link, will be seen pivoted at the base of the shaper and reaching up to the ram, to which it is connected by a short forked coupling. Behind the slotted link is a large gearwheel, which has on its face a short slide acting as a radius of the large gear. From this slide a crankpin projects and over it is fitted a rectangular block, which fits into the slot of the slotted link.*

If the machine is then turned by hand the crankpin block will be seen moving up and down the slot as the large gear rotates, at the same time imparting to the slotted link an oscillating motion, like an inverted pendulum. This in turn causes the ram to move forwards and backwards (a reciprocating motion).

The length of the ram stroke is controlled by the radius of the crankpin on the large gear. The short slide on which the crankpin is mounted is moved by a leadscrew, which is turned by bevel gears from a spindle through the bored centre of the large gear shaft. A detachable handle on the squared end of the spindle enables the adjustments to be made.

To locate the stroke evenly over the work, the clamp on top of the ram can be released to allow the coupling on the slotted link to take up any position within the slot in the ram.

Before starting a job the following adjustments have to be made.

When the tool and work have been secured the vertical slide has to be positioned to a convenient height relative to the tool, the slide locked and the front support adjusted and secured. The length of the stroke can then be determined by the scale on the ram slide and the position of the stroke by trial over the work allowing for stroke overlap. Finally the cutting speed can be set for the material being cut and the nature of the work to obtain the best conditions.

PROCESS 9 SHAPING

Object   To examine the problems of machining dovetail slides and the function of the clapper box, particularly on the 60° inverse face. Also to demonstrate a method of accurate measurement of the slide contact faces.

Material   Bright mild steel rectangular bar $30 \times 50 \times 75$ mm long. The end faces (30×50) to be machined square.

Tools and equipment   Shaping machine, machine vice, parallel strips as necessary, external micrometer (size according to dimension *A*, figure 9.1), standard Hoffman roller 20 mm diameter, depth micrometer, protractor, 150 mm steel rule, hide-faced hammer.

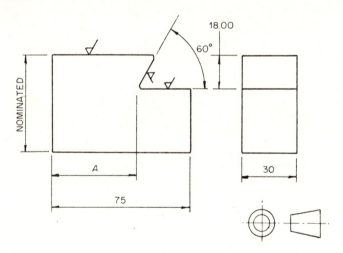

Fig 9.1

Instructions

Mark out position of dovetail from position given for *A*. This is to provide a guide for rough machining.

Machine Setting

Position vice on machine work table with jaws square to ram slide.

Set work in vice on parallel strips to allow top face of work to project 20 mm above vice jaws. Tap work down as vice is tightened. Fit tool in toolpost and set machine stroke to 40 mm. Position stroke to allow 5 mm overlap from each face.

Swing toolpost slide through 30° and secure. Release clapper box nut, incline to right as shown in figure 9.2 and secure. Adjust tool so that it will give clearance to all faces to be machined (figure 9.2).

Fig 9.2

Machining

Rough and finish top face. Rough out a square step by a series of cuts on cross-slide (figure 9.3). Rough out 60° face by a series of cuts on front slide (figure 9.4). Finish bottom face of dovetail to depth (18.00 mm) into corner and *note front slide index dial reading*. Calculate dimension over roller and take a trial finishing cut on 60° face. Measure over roller and repeat cuts until correct dimension is obtained (figure 9.5).

Remove sharp edges and have inspected.

Questions: Process 9

(1)   Assuming the front slide, the cross-slide and the ram slide to

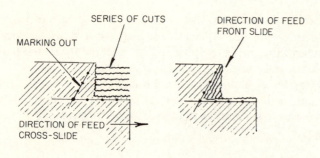

Fig 9.3

Fig 9.4

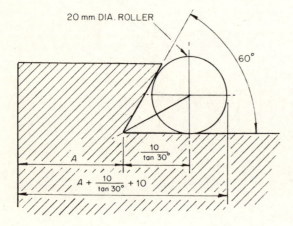

Fig 9.5

constant length in any position of the vertical slide.

(4)   What is the function of the clapper box?

(5)   In which direction, relative to a workface, should the clapper box be tilted?

be in a good state of repair, what would cause the tool to lift and be thrust sideways at the beginning of each cut and to leave a poor finish?

(2)   Sketch a cross-sectional view of a dovetail slide and indicate the bearing surfaces.

(3)   Sketch or describe how the centre distance between the cross-slide leadscrew and the ratchet feed crank wheel is maintained at a

Process 10

Marking-out, hand sawing and filing

Object

To show the potential of hand skills.

NOTES ON PROCESS 10: MARKING-OUT, HAND SAWING AND FILING

The type of work carried out in this process is particularly suited for marking-out, both in the time saved by unnecessary measurement during the manufacture and in its potential for reducing scrap work. In the checking procedure following marking, an incorrect line is easily put right but a sawn or filed dimension very often means scrapped work.

By raising the level of the work to a surface plate mounted on a marking-out table as shown in figure 10.2, the scriber of the surface gauge is more easily set in transferring dimensions from the rule to the work. Note that the rule and the work stand on a common base.

By first filing a datum base before marking-out and then using the finished sides of the material supplied, three datum faces are available for setting the work for marking-out. The advantage of this is shown in setting the work for the 45° lines.

It is a common fault in hand sawing to use excessive pressures on the saw. The blade is very thin and depends on the tension put into it from the frame to hold it straight. By using light pressures the blade will hold a straighter line and will give confidence to follow closely to the marking and reduce the amount of filing. The pitch of the teeth on the blade plays an important part in obtaining good results. Thin work like sheet metal or thin walled tubes requires closely pitched teeth so that there are two or more teeth bearing on the sawn face. In these circumstances the cutting action can be further improved by tilting the saw at an angle—in effect, cutting up hill. When sawing through thick material wider pitch teeth are necessary to give more clearance for the small particles of metal that tend to accumulate in the slot of the saw cut. All hacksaws are given clearance at the side of the blade by offsetting alternate teeth. When a saw blade tends to bind in the slot this is the first indication of wear and a loss of clearance.

When filing the vee faces it is better to reduce the two faces evenly, alternately changing the position of the work in the vice. It is essential that a hand safe edge file be used for this operation, that is, a file having one edge completely smooth. This design allows the file to work closely to the adjacent face of the vee while filing the other face. The difference between a safe edge file and a flat file should be noted:

the former has parallel edges and slightly tapering faces and the latter has tapering edges and parallel faces; if the two files are put side by side the difference can be clearly seen.

A three square file is necessary to clear the blurred corner left by the safe edge file and it needs to have fine cutting teeth to make a sharp corner. This file can also be used to put a finish on the completed work by draw filing. The finish of the uncut faces of the bright drawn stock bar should be left as supplied, since, provided that a smooth jaw vice or vice clamps have been used and maintained clear of swarf, the condition of the job will meet the requirements of the work.

PROCESS 10 MARKING-OUT, HAND SAWING AND FILING

Material   Bright mild steel flat bar $40 \times 6 \times 60$ mm long.
Tools and equipment   Marking-out table, small surface plate, surface gauge, rule and rule stand, square, a pair of medium-size vee-blocks, prick punch, light hammer, files: 200 and 100 mm safe edge, 100 mm fine three square, hacksaw fitted with medium-pitch blade, marking-out medium.

Instructions

Manufacture item shown in figure 10.1.

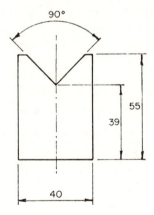

Fig 10.1

Preparation for Marking-out

File one end of the workpiece square to face and side. Remove burrs and sharp edges.

Marking-out

Cover one face with marking-out medium. Stand work on surface plate and support as shown in figure 10.2. Set surface gauge at 55 mm by rule and mark line (figure 10.2). Reset surface gauge to 39

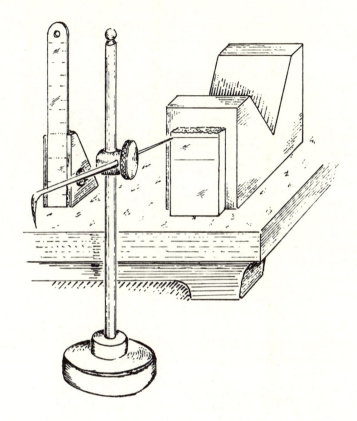

Fig 10.2

mm (bottom of vee), mark second line. Turn work on its side, set surface gauge to 20 mm, mark centre line (figure 10.3).

Place work in vee-block, support as shown in figure 10.4. Set surface gauge to intersection of second horizontal line to centre-line. Mark one side of vee. Turn work 90°, mark second side of vee. Lightly prick punch as shown in figure 10.5. (The order in which the lines are to be marked is given by the figures in this sketch.) The small dots will be cut in half when the work is finally finished to size.

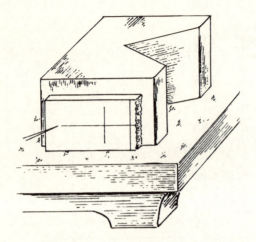

Fig 10.3

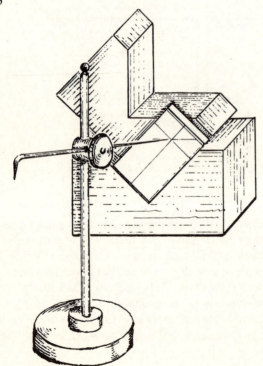

Fig 10.4

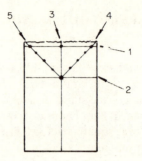

Fig 10.5

Sawing and Filing

Grip work in bench vice, using vice clamps as shown in figure 10.6. Position start of saw cut with three square file. Hand saw waste side of line. Reposition work in vice (figure 10.7); make a 'safe' start with file and saw to waste side of first vee side. Reset (figure 10.8) and repeat file and saw cut to remove vee waste.

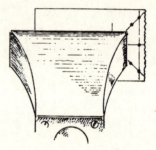

Fig 10.6

Rough out length and vee faces with 200 file, keeping safe edge to adjacent vee faces. File down to dots with 150 mm file, checking with square. Clear corner of vee with three square and draw file cut surfaces.

Remove sharp edges and burrs and check dimensions.

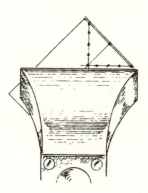

Fig 10.7

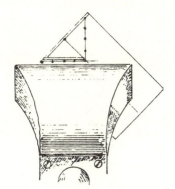

Fig 10.8

(3)   Name another form of supply that might have been used for this work.

(4)   Why is a hand safe edge file selected for filing the vee faces?

(5)   What effect does the pitch of the hacksaw blade teeth have on the thickness of the material being sawn?

Questions: Process 10

(1)   What is the purpose of filing one end of the workpiece square to the face and side before marking-out?

(2)   Suggest the advantages of using bright stock for this job.

Process 11

Marking-out

Object

To check the degree of accuracy obtainable by marking-out with a vernier height gauge, and to position the locations of machining areas, thus saving setting and machine loading times.

NOTES ON PROCESS 11: MARKING-OUT

Early experience in marking-out tends to destroy confidence in the accuracy of finished work that is machined to marking. If the job is properly done the results should be well within the general run of commercial limits. Much depends on how the work is prepared and the equipment used.

The work surfaces should be covered with a quick-drying marking-out paint, with the marked lines thin but clearly visible. A fine-pointed prick punch will ensure that dots intended to be positioned on a line are where they should be and not to one side or another—a common fault. Dots at line intersections, as in the case of a marked-out hole, require extra care, and easy though it may appear, results are too often unsatisfactory. The work in this process will provide an opportunity to see how effectively these measures have been carried out. A series of small dot marks, evenly spaced to show a machining line, is more likely to produce accuracy than large irregular indentations made by a poorly ground centre punch.

All marking should be carried out on a marking-off table, which ought to have a good range of equipment such as vee-blocks, rule stand, jacks, parallel strips, wedges, angle plates and adjustable centres, as well as a standard range of small tools, gauges and measuring instruments. If the working surface can be raised on a secure stand to about 1.2 m, the marking can be carried out without excessive stooping and vernier scales can be read much more conveniently. The area around the table must be free from obstructions and stand in a good light.

The vernier height gauge is an ideal instrument for marking-out where the standard of work demands precision. The thickness of a marked line should not be overlooked and for this reason the vernier height gauge 'nib' should be kept sharp and free from 'rags' left from grinding. It is reasonable to expect a marked line to be from 0.10 to 0.15 mm thick.

When angular settings are required the combination set protractor or the vernier protractor can be used, but this involves setting a line to the thick blade of the protractor, which can lead to parallax errors. A dividing head fitted with a base plate to stand on the marking-off table can very often simplify this kind of work if it can be adapted for mounting in a chuck or on a mandrel. Given the normal ratio of 40 to 1, one turn of the crank equals 9° on the work, further divisions being obtainable on the hole circles of the index plate.

PROCESS 11 MARKING-OUT

Object To demonstrate marking-out with a vernier height gauge and to position the locations of machining areas.

Material Mild steel bright round bar 50 mm diameter × 90 mm long. *Note* The bar must be free of burrs on the circumference.

Tools and equipment Centre lathe, three-jaw chuck and right-hand facing tool, surface plate, vernier height gauge, vee-block that will permit sides of work to be marked as well as faces (see figure 11.2), sharp centre punch, 150 g hammer, 100 mm dividers, engineers' square, set of feeler gauges, marking medium.

Instructions

Face ends of bar to length and remove sharp edges.

Prepare for marking-out work shown in figure 11.1.

Set work in vee-block on surface plate together with vernier height gauge. 'Touch' top of work with 'nib' of vernier height gauge testing with thin blade of feeler gauge (figure 11.2). Note vernier scale reading.

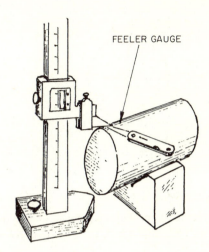

Fig 11.2

reading and mark centre-line all round work. Do not use heavy pressures or try to score deep lines.

Mark once only.

Marking Second Centre-line

Turn bar 90° to bring first centre-line vertical. Check with square (figure 11.3). Mark second centre-line as for first. Both ends of bar should be as shown in figure 11.4.

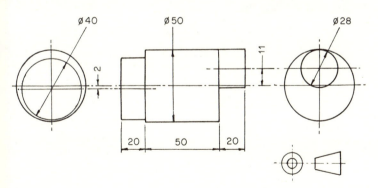

Fig 11.1

Marking the Work Centre-line

From vernier reading obtained above, subtract half diameter of bar (25 mm) plus feeler gauge thickness. Reset vernier to new

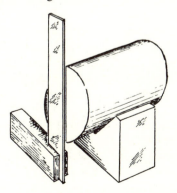

Fig 11.3

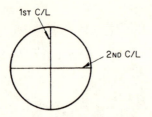

Fig 11.4

Marking for the Eccentric Centre-lines

Raise slide of vernier from centre setting 11.00 mm (see right-hand sketch of figure 11.5) and mark centre-line of 28 mm diameter eccentric.

Lower slide 11.00 + 2.00 mm to mark centre-line of 40.00 mm diameter eccentric on opposite face. End faces should be as shown in the two sketches in figure 11.5.

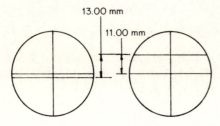

Fig 11.5

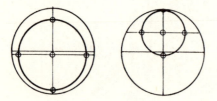

Fig 11.6

To Complete Marking-out

Centre-dot centres of eccentrics, mark eccentric diameters with dividers and again dot to make position 'permanent' (figure 11.6). *Note* The 'dots' are exaggerated to stand out clearly in the sketch.

Questions: Process 11

(1)   Name three essential requirements of a good marking-out medium.

(2)   What is meant by the term 'making the marked positions permanent'?

(3)   Suggest another type of vee-block that would prevent any tendency for the work to tilt during marking-out.

(4)   Where is the datum face of the vernier height gauge?

(5)   What will be the effect of this marking-out on machine setting and loading time?

Process 12

Spacing holes on a pitch circle

Object

To show the potential of hand skills: power hand drilling, marking-out.

NOTES ON PROCESS 12: SPACING HOLES ON A PITCH CIRCLE

Before using electrically powered hand tools an examination should be made of the cable for cuts and abrasions that could lead to a failure of the insulation. Loose connections at the plug and at the entry of the cable to the drill, as far as they are visible from the outside, should be reported to the department responsible.

Long leads between the power socket and the work are a potential source of danger. If possible they should be carried on supports above head height with suitable notices above gangways to warn personnel carrying ladders or high loads. (See pp. 6–7, points 1 to 7.)

The procedure for drilling the 6 mm holes is designed to give the best results for accuracy in positioning and in hole diameter. Different methods would be adopted in production or where the problems involved were not under examination or study.

The 'point' of a twist drill has a chisel-shaped cutting edge and this feature increases with the drill diameter. The 3 mm drill has a small chisel point, which will readily locate itself into the marked centre dot. The drill need only penetrate the work sufficiently for the following drill to take up the correct alignment. If a drill is ground with one cutting edge longer than the other, or at different angles to each other, the drill will produce a hole larger than the drill, the reason being that the drill rotating on its point cuts a hole twice the radius of its longer cutting edge, approximately.

In the general run of workshops, drills of this size are commonly ground offhand and it requires some skill and experience to grind correctly. For this reason the second drill, 5.5 mm, provides a margin of error. The final drill, 6.0 mm, has no solid material at its centre for the point to rotate on, and provided it is reasonably ground it will 'scrape' the hole correctly to size.

The block of wood behind the work held in the vice stops the drill tearing through the plate when the chisel point collapses at the breakthrough. The right-hand spiral of the drill flutes acts in a rapid screw action. In addition, without the resistance of the wood block, the operator, possibly using excess pressure, will suddenly be thrown off balance and on to the work.

Marking out holes on a pitch circle is a frequent requirement in a general engineering workshop, as, for example, in a bolted pipe flange. In this case the work is more difficult, since there is no centre for the dividers to mark the circle. The usual procedure in this situation is to block the centre of the pipe with a piece of wood or in the case of a large bore to bridge it with a suitable flat bar chamfered at the ends and wedged in the bore mouth. The centre can then be found by odd leg calipers either from the bore or the outside diameter of the flange.

The spacing of four or six holes is easily stepped out on the pitch circle but the equal division of five or seven holes is a little more difficult. Probably the most convenient method would be to mark the pitch circle on scrap sheet metal and with a protractor mark two lines from the centre to the circumference at an arc equal to the divisions required. For example, 5 spaces $= 360°/5 = 72°$. The chordal distance can then be set on the dividers and transferred to the work.

PROCESS 12 SPACING HOLES ON A PITCH CIRCLE

Object   Power hand drilling and marking-out.
Material   Bright mild steel flat bar $100 \times 3 \times 103$ mm long.
Tools and equipment   Bench vice with soft jaw clamps, block of hard wood about the size of workpiece 25 mm thick, hand power drill 6 mm capacity, marking-out table, angle plate, G-clamp, surface gauge, rule and stand, engineers' square, centre punch, hammer, hacksaw with fine tooth blade, files: 200 mm flat second cut, 150 mm hand fine.

Instructions

Manufacture item shown in figure 12.1.

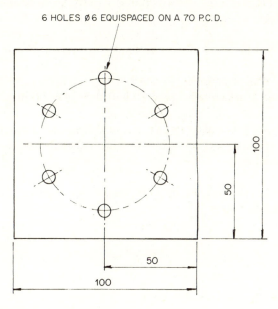

6 HOLES Ø 6 EQUISPACED ON A 70 P.C.D.

Fig 12.1

Preparation
File one of the saw cut edges flat and square to the face and side check with engineers' square. Remove burrs and sharp edges.

Cover one face with marking medium.

Marking-out

Set work on marking-out table, filed edge as base, support with angle plate, secure with G-clamp (figure 12.2).

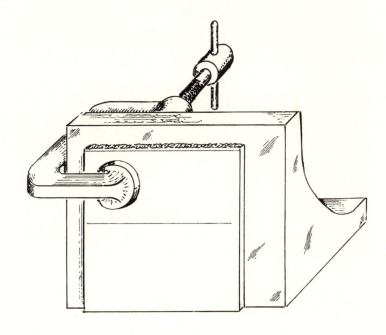

Fig 12.2

Set scriber of surface gauge to rule at 100 mm, mark line. Reset scriber to 50 mm, mark centre-line. Release work, turn on to adjacent edge (90°), reclamp, mark second centre-line.

Unclamp work, lay flat, centre dot where centre-lines cross. Set dividers to 35 mm radius, mark pitch circle. Centre dot at opposite ends of *one* centre-line where pitch circle intersects centre line (figure 12.3). Recheck dividers at 35 mm, mark arcs on pitch circle from opposite ends of centre-line.

Centre dot remaining four arcs at intersection with circle.

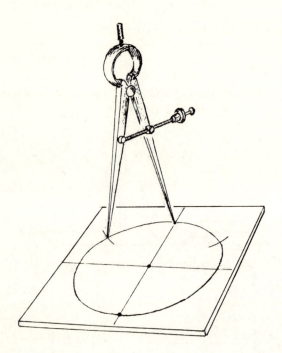

Fig 12.3

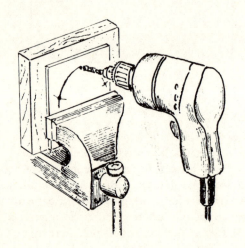

Fig 12.4

Drilling

Select a bench vice near a power point. Grip work with wood block backing (figure 12.4). Fit 3 mm drill to power drill chuck, place power plug in socket, drill the three exposed centre punch holes. Reset work for remaining three holes. Remove 3 mm drill, replace with 5.5 mm drill. Repeat former drilling procedure. Finally open up holes with 6 mm drill. Exert light pressures only when drilling. Remove 'rags' left by drilling with smooth file. Check hole spacing by measuring from edge of one hole to edge of next, check diametrically in same way.

Question: Process 12

(1)   (a) What is the purpose of marking-out from two opposite sides of the centre-line? (b) Name two possible causes of error arising from marking-out.

(2)   Why drill 5.5 mm first and finally to size with 6 mm drill?

(3)   (a) What is meant by the chordal distance between the holes? (b) For six holes equally spaced what is the chordal distance?

(4)   By simple line diagrams show how the chordal spacing of the drilled holes can be checked by using the 'outside of the jaws' of a vernier caliper.

(5)   Considering the use of power hand tools, outline five basic safety requirements.

Process 13

Marking-out, drilling and reaming

Object

To show the potential of hand skills: spacing holes accurately, obtaining a good finish and close dimensioning of hole diameter.

NOTES ON PROCESS 13: MARKING-OUT, DRILLING AND REAMING

Marking-out with a vernier height gauge is necessary when a greater degree of accuracy is required than is obtainable by a surface gauge and rule. Since the base of the vernier height gauge is one of its measuring faces, it is an advantage if the work rests on the same datum face, as shown in figure 13.2.

To place the centre-line exactly in the centre of the work and to compensate for allowable variations in the bright stock width, it is necessary to obtain a 'touch' reading on the vernier scale. This is done by interposing a thin feeler blade between the 'nib' of the vernier height gauge until the 'touch' can be felt by moving the blade. From the reading obtained, subtract the thickness of the feeler to give the true width of the work, or height as it will be in this case. By halving this dimension the centre-line can then be set.

In the second work setting—spacing the hole centres—there is no common datum base for the vernier height gauge and the work. A datum dimension will therefore have to be taken from the top end of the work and this is obtained in the same way as the setting for the width centre-line.

The advantage of good-quality marking-out equipment for the surface table is demonstrated in the angle plate holding the work. By turning the angle plate on its side the work is automatically set at right-angles to its former setting. A box angle plate serves the same function.

The sensitive drilling machine used for this work is one where the pressure on the drill can be felt through the spindle operating handle. This is of great value when drilling very small holes. A simple rack-and-pinion mechanism is used. The rack is cut on the outside of a sleeve through which the spindle passes and rotates within it. At the lower end of the sleeve a ball thrust race takes up the thrust on the drill. A pinion, which engages with the rack, is mounted on the operating handle spindle and turns with it. A clock-type spring on the end of the pinion spindle ensures that the drill returns to its uppermost position when the handle is released. This removes the danger of the drill falling on the work during setting on the drilling machine work table or by an accidental release of the handle. Most machines are fitted with drill depth indicators or depth stops. A

sensitive drilling machine is designed for the smaller sizes of drill and operates at high spindle speeds. It is usual for the range of spindle speeds to be obtained by means of a pair of cone stepped pulleys. The power of the motor pulley is transmitted to the spindle pulley by means of an endless vee-belt. To effect a change in spindle speeds a guard has to be lifted and the tension of the belt relieved by a cam action on the motor mounting. The driving belt may then be changed to another pair of pulleys. For low drilling speeds the small pulley on the motor should drive a large pulley on the drill spindle. Since the cone stepped pulleys are the same size, only mounted in reverse, the driving belt will fit any pair of opposite pulleys.

Reaming is an operation designed to produce holes with a good finish and to closely controlled sizes. There are hand reamers and machine reamers, the latter usually have taper shanks to fit machine spindles. Their cutting action is to scrape rather than remove bulk stock and for this reason the amount left in a hole to be reamed has to be quite small—just sufficient to clean up is ideal. The forward end of the reamer is slightly tapered for about one-quarter of its length, which helps in the alignment to the hole and in the progressive removal of material. The best results are obtained at low cutting speeds and at comparatively fast feeds coupled with a generous flow of cutting lubricant.

Expansion and expanding reamers are available for holes between the standard sizes. The former are expanded by an internal tapered pin, which causes the reamer to bulge outwards from the bored and slotted centre. The range of expansion is very limited. The latter operate on a different principle: the body of the reamer is taper slotted throughout its length and is externally threaded. Matching stiff blades fit into the slots and are retained by internally threaded and chamfered sleeves, which bear on the ends of the blades and hold them in position. The size of the reamer is controlled by moving the blades along their tapered slots by means of the sleeves at each end of the blades. The range of sizes possible from one reamer is very useful. The outsides of the reamer blades are, of course, ground parallel.

PROCESS 13 MARKING-OUT, DRILLING AND REAMING

Object   Spacing two holes accurately, obtaining a good finish and close dimensioning of hole diameter (figure 13.1).

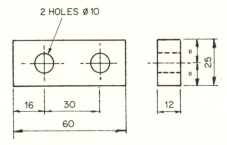

2 HOLES Ø 10

16   30   60   25   12

Fig 13.1

Material   Bright mild steel flat bar $25 \times 12 \times 60$ mm long, one end machined square to sides.

Tools and equipment   Marking-out table, angle plate, toolmakers' clamp, vernier height gauge, centre punch, hammer, marking medium, sensitive pillar-type drilling machine, drill chuck, drilling vice, centre drill, twist drills 8.5 and 9.5 mm diameters, machine reamer 10 mm diameter vernier caliper, parallel strips, dividers.

Instructions

Remove all burrs and sharp edges from work. Apply marking medium on surfaces to be marked-out.

Marking-out

Clamp work against angle plate as in figure 13.2, making sure edge is located firmly on to marking-out table.

Touch 'nib' of vernier height gauge on top surface of work (figure 13.2). Record reading.

Height gauge reading ÷ 2 = new setting. Reset height gauge. Mark centre-line (figure 13.3).

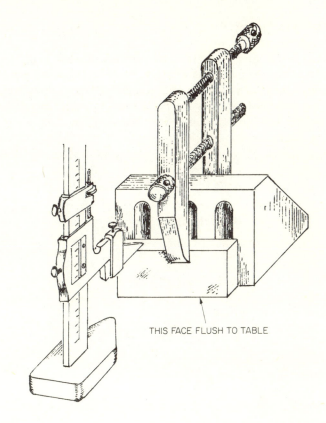

THIS FACE FLUSH TO TABLE

Fig 13.2

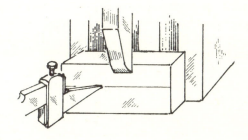

Fig 13.3

Rotate angle plate 90°, centre-line is now vertical and machined end uppermost.

Reset height gauge 'nib' to touch top edge of work as in figure 13.2. Record reading. Deduct 16 mm. Reset height gauge, mark a line 90° to centre-line (figure 13.4). Deduct 30 mm from previous reading and reset. Mark a second line 90° to centre-line (figure 13.4). Use only light pressures while marking lines.

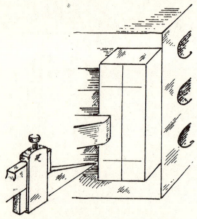

Fig 13.4

To Complete Marking-out

Unclamp work from angle plate. Lightly centre punch hole centres. Set dividers and mark 10 mm circles (figure 13.5). Centre dot at centre-line intersections to circle. *Note*   The centre punch holes are exaggerated in the diagram.

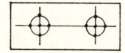

Fig 13.5

Prepare for Drilling

Clamp work in drilling vice. Ensure positive location on parallel strips (figure 13.6).

Fit drill chuck to machine spindle, secure centre drill to chuck. Position vice below machine spindle. Select correct spindle speed and start machine.

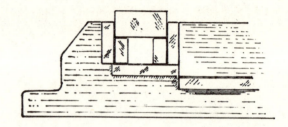

Fig 13.6

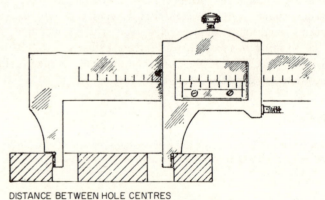

DISTANCE BETWEEN HOLE CENTRES
= (VERNIER READING + JAW THICKNESS) −10 mm

Fig 13.7

Drilling

Using extreme care align centre drill with marking of first hole.

Centre drill. *Note*   At this point it is advisable to clamp vice to machine table.

Remove centre drill. Adjust spindle speed to suit 8.5 mm drill. Drill. Open hole out to 9.5 mm diameter. Reduce spindle speed and ream to size (10 mm). Ample cutting fluid should be used throughout machining operations.

Repeat the above drilling operations for second hole. Remove work from vice and deburr.

Use outside measuring 'nibs' on vernier caliper gauge to measure maximum distance from one hole to another (figure 13.7). Measurement should be 40 mm.

Questions: Process 13

(1)   (a) Give two reasons why it is advisable to clamp the vice to the machine table. (b) Why is it necessary to drill the 9.5 mm hole?

(2)   What is the reason for reducing the spindle speed prior to the reaming operation?

(3)   Considering that the direction of rotation is the same, briefly explain why a drill has a right-hand helix and a reamer a left-hand helix.

(4)   Apart from the toughness or hardness of the material being drilled what is the cause of the drill 'burning' at its cutting edges?

(5)   Explain using a sketch in your answer another way of checking the centre distance of the holes.

Process 14

Marking-out, heat treatment, drilling and bending

Object

To examine the mechanical properties of mild steel; the application of powered hand tools.

NOTES ON PROCESS 14: MARKING-OUT, HEAT TREATMENT, DRILLING AND BENDING

The 'form of supply' of the material given for this work greatly simplifies the manufacture. There is very little stock to be removed and this only on the ends of the bracket 'legs'.

Designers, in the interest of economy, have to keep in mind the standard 'forms of supply' (that is, the sectional shape of the material) that are available at major metal stockists. They are also influenced by the numbers required and must consider the method of manufacture. For one-off there are no particular problems except perhaps the strength and finish.

Bright stock, by being finally cold drawn or cold rolled to bring it accurately to stock dimensions, have stresses set up on and near the surface. These stresses must be relieved before bending to avoid cracks developing on the outside of the bend; hence the early heat treatment operation on this work. Black rolled bar does not have this property to the same degree and might be considered more suited for this work, except that the dimensions of the stock are not so closely controlled and the corners and sides tend to be rounded and the finish matt and scaled.

Where large numbers are required the method of manufacture would be quite different from that given in the work schedule. The holes and bending would be done on a two-stage press and the time to make one would be a matter of a few seconds, but the quantity would have to justify the costly manufacture of suitable press tools. With just one-off or for a small job lot, the method given in the schedule ensures an acceptable degree of uniformity.

The main problem in making one is in the positioning of the bend for the right relationship to the holes. Therefore, by bending first in the approximate centre of the stock and having the surplus length on the ends of the bracket, the marking-out follows naturally the dimensioning on the drawing, except that, if the marking is carried out on the outer faces of the bracket instead of the inside faces as shown on the drawing, the holes must be positioned on opposite sides. In this way the drilled holes will 'hand' the bracket (that is right and left hand). The wrong 'hand' may mean a failure to match a joining component.

The bending operation should be done with the least number of

blows possible using a hide- or copper-faced hammer. Alternatively a blacksmith's flatter might be used to advantage, holding the flatter face squarely on the part to be bent and hitting the striking face with a fairly heavy hammer. Too many blows cause unnecessary stresses in the work and tend to bend up the end of the bracket.

A close examination of the distortion at the bend should be made to see the effect made on the shape of the material section. The outside of the bend will be seen to have stretched, while on the inside there is compression causing the metal to 'flow' towards the edges, giving a distinctive pointed projection that has to be filed off before marking-out from the side. The thickness at the bend will be seen to have thinned.

The brazing hearth used for normalising this work is usually made up of a cylindrical shallow container about 1 m diameter, lined with firebrick and mounted on a stand to bring it to a convenient working height. Town or bottle gas is fed to the blowpipe by means of an armoured flexible pipe and a forced air supply from an electrically driven compressor mounted on the underside of the hearth. The air supply is also connected to the blowtorch by a flexible pipe. Two thumb lever operated cocks regulate the flow of air and gas, which in turn control the flame and temperature. This method of normalising leaves a hard scale owing to oxidation, and cleaner work can be obtained from temperature-controlled electric furnaces where the air is excluded.

PROCESS 14 MARKING-OUT, HEAT TREATMENT, DRILLING AND BENDING

Object   To examine the mechanical properties of mild steel and the application of powered hand tools.

Material   Flat bright mild steel $40 \times 3 \times 95$ mm long.

Tools and equipment   Marking-out table, scriber, surface gauge, centre punch, 300 mm rule and stand, marking medium, 3 mm radius gauge, 6 mm diameter drill, engineers' square, 200 and 100 mm hand files, hand hammer, bench vice with smooth jaws, brazing hearth equipped with gas torch, bending block (figure 14.3) with 3 mm radius on one edge, electric pistol drill capacity 10 mm, 6 mm drill, copper/hide-faced hammer.

Instructions

Manufacture item shown in figure 14.1. Remove all burrs and sharp edges from work.

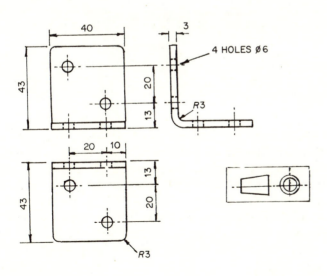

Fig 14.1

Preparation prior to Marking-out—Stress Relieve

Light gas torch and adjust to give a moderately fierce flame. Place work conveniently on brazing hearth and direct flame into work. Heat until work appears dull red.

Switch off gas torch. Use tongs to remove work from hearth and allow material to cool in air. *Note*   Because of the danger of another student being burnt by picking up hot work, place the work out of reach of other people.

Marking-out for Bending

Apply marking medium to one surface of work. With rule, scriber and engineers' square establish a centre-line as shown in figure 14.2.

Fig 14.2

Bending

Place work and bending block in vice as shown in figure 14.3, making sure that scribed line is parallel with point of contact with block. Using copper/hide-faced hammer form work round prepared 3 mm radius. Test with radius gauge.

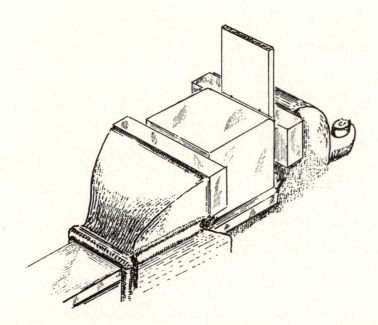

Fig 14.3

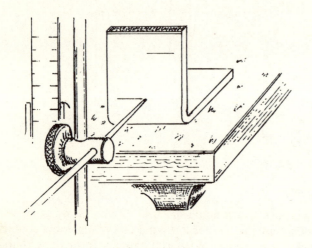

Fig 14.4

Complete Marking-out

Remove marking medium remaining from bending operation and reapply medium to surfaces to be marked (figure 14.4). Position work on surface plate.

Set scriber to 13 mm, mark line, turn bracket over to mark second line at 13 mm on adjacent face.

Reset scriber to 33 mm, mark second set of centre-lines on both faces. Reset scriber to 43 mm to mark over-all length of bracket faces (figures 14.4 and 14.5).

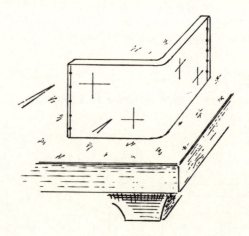

Fig 14.5

Reposition bracket on its side (figure 14.5). Set scriber to 10 mm, mark hole centre-line on both faces. Reset scriber to 30 mm to mark second set of centre-lines on both faces.

Prick punch over-all length lines, centre punch centre-line intersections of hole centres (figure 14.5). *Note*   Marking on outer faces of bracket will be reversed to the positions shown in figure 14.1.

Drilling Holes

Fit 6 mm drill in pistol drill chuck. Plug lead into nearby socket.

Note   Observe safety features outlined on pp. 6–7. Place work and packing in bench vice as shown in figure 14.6. Position point of drill in centre dot.

Check alignment of drill axis is 90° to surface of work (figure 14.6). Squeeze starting switch. Apply pressure to drill allowing it to cut and continue to drill hole through work into wooden packing piece. Switch off.

Repeat drilling operations for remaining holes.

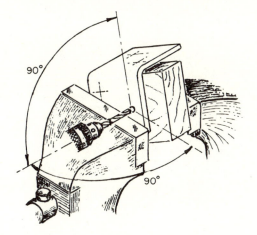

Fig 14.6

Filing

Remove 'rags' left by drilling. Use wooden packing block as backing for inner face, hacksaw and file bracket legs to length, working to marking (figure 14.5).

File 3 mm radius on each corner of bracket (figure 14.1). Test with radius gauge.

Remove sharp edges and clean up.

Questions: Process 14

(1)   If all the filing and drilling were done before bending what would be the developed length of the work?

(2)   (a) Explain what happens to the inner and outer fibres of the metal at the bend. (b) What is meant by (i) bending across the grain and (ii) bending with the grain?

(3)   (a) Suggest an alternative 'form of supply' that would not involve bending. (b) Would this be stronger?

(4)   Why is it important to use a packing piece to support the work while drilling?

(5)   (a) Name the heat treatment process prior to bending. (b) Why is the heat treatment necessary? (c) What effect does the heat treatment have on the mechanical properties of the steel?

Process 15

Accurate hole spacing

Object

To study the function and capability of a compound table fitted to a standard drilling machine for jig boring.

NOTES ON PROCESS 15: ACCURATE HOLE SPACING

A compound table is made up of two hand-scraped or ground slides set at right-angles to each other in a robust casting. The movement of the slides is controlled by accurately machined leadscrews, fitted with indexing dials for registering slide movement. In addition, vee-platforms, parallel to the slides, are provided to enable length bars and slip gauges to rest on them to act as spacers between a faced lug on the side of the slide and a dial indicator mounted on the platform. Thus the movement of the slides can be determined by two methods: by the leadscrew and by the dial indicator; the latter will give the best results.

The compound table is a very useful item of ancillary equipment, adaptable to milling and drilling machines, although it is primarily intended for drilling machines. A good-quality pillar-type drilling machine with suitable power feeds and sound spindle bearings virtually becomes a jig borer. It cannot, of course, compare with the all-round quality of a jig borer or cope with the same range of work, but it can produce a high standard of work within the limited movement of the slides.

Since the workpiece supplied has two adjacent edges ground square, it is only necessary to true one edge to a slide, preferably the longer YY datum edge. The shorter XX edge will automatically line up to the second slide.

Considerable care is necessary in setting the spindle test plug to the datum edges of the work. The method is explained and illustrated in the work sheet. A light touch with the feeler blade between the test plug and datum edge is essential to avoid deflecting the drill spindle, and a further test should be made after turning the spindle.

The object of using a boring head after drilling is to correct any misalignment the drill may have taken at the beginning of the cut. The amount of error will be small but needs to be corrected by the tool of the boring head rotating on the true axis of the spindle. This has the additional advantage of opening up the standard-size drilled hole to an ideal reaming size.

Where there are several drill changes for every hole, the quick-change drill chuck might be used to advantage. This type of chuck has a number of special adaptors, which are internally bored to a standard Morse taper. Externally they are designed to fit into the chuck and are locked by a single upward thrust of the adaptor into the chuck and released by a similar motion of a freely rotating knurled ring. Both fitting and releasing can be carried out while the drill spindle is in motion.

As a check against the possibility of large errors in calculating the rectangular coordinates, the work may be marked out using polar coordinates. The marking will serve as a guide and give confidence while carrying out the work.

PROCESS 15 ACCURATE HOLE SPACING

Object   To demonstrate an application of a compound table fitted to a standard drilling machine for jig boring.

Material   Bright mild steel bar $100 \times 10 \times 120$ mm long, nominal dimensions. The two datum edges XX and YY to be ground square to each other and to the plate face.

Tools and equipment   Vertical drilling machine, compound table (figure 15.1), boring head (figure 15.2), drill chuck, centre drill, taper shank drill 14.5 mm diameter, machine reamer 15.0 mm diameter, set of slip gauges with accessories (figure 15.3), 0–25 mm micrometer, small telescopic gauge, 150 mm steel rule, d.t.i. with drill spindle mounting (figure 15.4), parallel strips, suitable clamps to secure work to compound table.

Secure compound table to drilling machine so that drill spindle is approximately central to slides. Lightly clamp work over parallel strips so that even pressure is applied to plate. Fit d.t.i. to drill spindle (figure 15.4) and line up YY datum edge to top slide. Finally clamp work. Drill, bore and ream holes spaced as shown in figure 5.5.

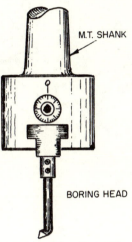

Fig 15.2

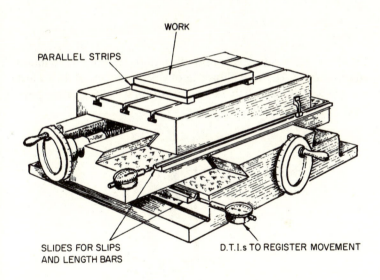

Fig 15.1

Instructions

Convert polar coordinates given in figure 15.5 to rectangular coordinates shown in figure 15.6.

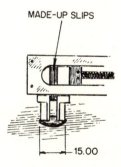

Fig 15.3

Fig 15.4

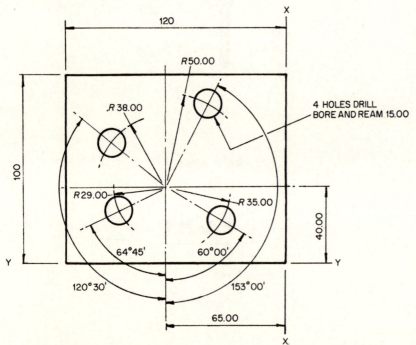

Fig 15.5

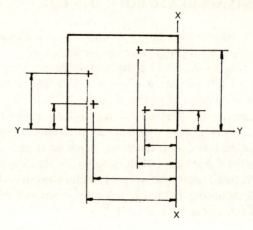

Fig 15.6

Setting

Fit Morse taper plug to drill spindle and bring test plug to datum edge (figure 15.7). Raise plug clear of work and move slide towards spindle centre-line, half diameter of plug plus feeler gauge thickness.

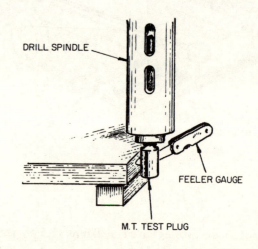

Fig 15.7

Reset indexing dial to zero. Repeat for second datum edge. Note that in all these and subsequent final setting operations, slides must be fed in *one* direction only to keep any slackness in leadscrew nut and thrust bearings acting one way.

Machining

Starting with hole nearest to two datum edges (bottom right, figure 15.5) fit drill chuck and centre drill and index slides to calculated coordinates. Clamp both slides, centre drill, drill 14.5, bore 14.8 ream and deburr. Complete remaining three holes in same order.

An alternative and more sensitive method of registering slide movement can be obtained by using the dial indicators shown in figure 15.1. These form part of the kit supplied with the compound slide.

The auxiliary slides also shown in figure 15.1 provide a base for slip gauges and length bars, which can be 'made up' for zero setting and then 'added to' or 'taken from' to obtain coordinate dimensions.

Questions: Process 15

(1) (a) What is the reason for boring after drilling? (b) What is the advantage of reaming after boring?

(2) Why is it necessary to clamp the work on parallel strips instead of directly on the compound work table where it would be better supported?

(3) Describe a method of checking the positioning of the holes to the accuracy required.

(4) In the work setting instructions, provision is made for lining up the YY edge only to the top slide of the compound table. Why are similar instructions omitted for the XX edge?

(5) How are the table slides secured after each hole location against possible movement by vibration and tool thrust?

Process 16

Soft soldering

Object

To study the nature of a soft soldered joint and other forms of soldering.

NOTES ON PROCESS 16: SOLDERING

Soldering is a process for joining metals, in the same way that wood and some plastics are joined with an adhesive compound. There are, in the main, two types of soldering: soft soldering and hard soldering. The choice of either type depends on the nature of the joint to be made and the strength required.

All soldering requires the application of heat, either by soldering iron, blowtorch or by dipping the work in a container of molten solder; the latter (called tinning) is applicable to the higher-tin-content soft solders.

The surfaces to be joined must be free from grease and scale for the solder to 'take', but at the same time the heat required to melt the solder causes oxides (a changing colour film) to form on the cleaned surfaces. To stop this occurring a flux is used to provide a shield against the action of the oxygen in the atmosphere on the heated surfaces.

Types of Solder

Soft Solder

Generally soft solder is an alloy of tin and lead. The higher-tin-content solders have a lower melting temperature and flow more freely than the lead solders.

Solder with a tin/lead content of about 60/40 is known as tinman's solder, which melts at about 183 °C. This type is used for electrical, radio and instrument assemblies. By a slight increase in the lead content the melting temperature is raised a few degrees, making it suitable for less sensitive and lower-cost work in the tinsmith's workshop. A marked feature of tinman's solder is its rapid transition from liquid to solid and vice versa.

Plumber's solder has a tin/lead content of about 40/60 and has a melting temperature of approximately 250 °C. It is used for joining lead pipes or lead sheathing of electric cables. Unlike tinman's solder the transition from liquid to solid is comparatively slow, providing a pasty stage, which allows the solder to be shaped or 'wiped' by a suitably gloved hand.

Hard Solder

Hard solder has a higher melting temperature and is considerably stronger than soft solder.

Silver Solder

Silver solder has an average composition of 34 per cent copper, 50 per cent silver and 16 per cent zinc. It is used for making a strong joint in small parts unsuited to the higher temperatures of brazing. The melting temperature is about 600 °C. The high silver content makes it rather expensive, which limits its use in large assemblies. It is usually applied to the work with a small blowtorch.

Brazing Solder

Brazing solder is a copper/zinc alloy in an average proportion of 60/40. It has a melting temperature of 850 °C and is available in granular form, wire and thin rods. Owing to the high melting temperature it is confined to making joints in ferrous material. One particularly useful application is in securing carbide tips to the shanks of lathe tools; some inserted tooth milling cutters are similarly secured. Brazing has a universal application but it is slowly being superseded by the improved techniques of welding.

Fluxes

There are two types of flux for soft soldering: an acid known as 'killed spirits', which both cleans the surfaces and at the same time provides a shield against oxidation; and a passive flux, which excludes the air during the soldering process.

The acid (active) flux is made by dissolving zinc in hydrochloric acid. It is highly corrosive and unsuitable for electrical and radio components but it is used extensively in tin and copper work. The parts must be washed thoroughly when completed.

Passive fluxes have a resin base made up in the form of a paste, which is spread upon the work prior to soldering. It has no corrosive action.

Flux used for silver soldering and brazing is usually in the form of a powder known as borax. As the borax approaches brazing temperature it melts and has a solvent action on the oxide, thus cleaning the metal and allowing the solder to unite with the work.

As the practical work in this process demonstrates, the adhesion of the solder to the parent metal is stronger than the solder that joins them; this applies to all forms of soldering.

PROCESS 16 SOFT SOLDERING

Object To study the nature of a soft soldered joint.
Material Two pieces of tin plate 0.48 mm thick, one 10 × 55 mm, one 14 × 55 mm.
Tools and equipment Soldering-iron oven, medium-weight soldering iron, stick of tinman's solder, resin-based soldering flux, flat-nose pliers, clean flat firebrick, medium emery paper (to clean soldering iron), wood dowel about 300 mm long to hold faces together when hot.

Instructions

Solder the two tin plate strips together as shown in figure 16.1.

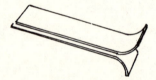

Fig 16.1

Preparation

Gather workpieces, tools and equipment conveniently close to oven so that there is little heat loss in transferring iron from oven to work. Bend ends of strips for a quarter of length as shown in figure 16.1, keeping remainder flat. Cover one face of each strip with thin layer of flux. Lay strips on firebrick, flux uppermost. Clean tapered end of soldering iron, light oven and place iron on top shelf.

Soldering

Watch for beginning of a green/blue flame coming off iron. When this shows, remove from oven, dip cleaned end into flux and immediately test on solder stick. If solder runs freely to iron, correct temperature has been reached. Cover 10 × 55 mm strip with thin layer of solder, lift strip with pliers and place over 14 × 55 mm strip.

Reheat iron but do not overheat.

Reflux iron and lay on assembled strips (figure 16.2). When solder is seen to run freely round sides of narrow strip, hold pieces together with dowel, lift off iron. Molten solder will dull slightly when it solidifies.

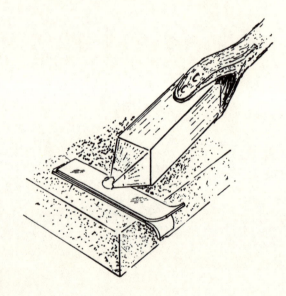

Fig 16.2

Testing and Examining the Joint

Hold one bent end in vice, grip other end in pliers. Separate by a steady pull (figure 16.3). Exposed faces should be a bright grey colour uniformly spread over whole split area. If these conditions have not been obtained work should be repeated with two new pieces of tin plate.

Probable causes of failure: insufficient heat, joining faces not clean, insufficient flux.

Fig 16.3

Questions: Process 16

(1)   In the test carried out how did the pieces separate: (a) by the solder leaving the parent tin plate, or (b) by the solder dividing leaving an even deposit on each face?

(2)   What are the physical indications that the soldering iron has reached the right temperature for making the joint?

(3)   Why is copper used for the soldering 'bit'?

(4)   What is the purpose of using a flux?

(5)   When would an 'active' flux be used in preference to a 'passive' flux?

Process 17

Milling—producing a concentric square on a round bar. (Figure 17.1)

Object

To detail a method of ensuring positive concentricity of the square to the round bar.

NOTES ON PROCESS 17: MILLING

Deviations in concentricity between the bar axis and the completed square are reduced by the method described in the process sheet. Errors incurred when touching the cutter on the circumference of the bar (either heavy or light touches) will result in variations between one flat to the centre, but by following the procedure of taking a trial cut, measuring and then indexing to final size for each flat, more consistent results will be obtained.

The principles of down cut *milling have the advantages of providing greater rigidity by directing cutting forces downwards on to the work table and leaving a better surface finish on the work. However, unless the machine is equipped with a backlash-eliminating device, only* up cut *milling should be used. This is because with the down cut technique working clearances between leadscrew and nut are immediately taken up by the cutter as it touches the work. It snatches at the workpiece, drastically increasing the depth of cut and from then on trying to drive the machine table along, causing the cutter to ride over the work. This leads to tool breakage, spoilt work and a bent arbor. Hence the process is sometimes referred to as 'climb' or 'grab milling'.*

It is worth while observing the application of the key fitted to the arbor. Its main function is to lock the cutter firmly on the arbor during cutting giving a positive drive. With no key fitted the drive is provided by friction between the cutter and the collars on either side of cutter, and for most applications this is sufficient for satisfactory results. Omitting it can serve as a useful safety device by allowing the cutter to slip on the arbor, thus, in an emergency, preventing expensive cutter breakage.

Under the base of most machine vices there is provision for two tenons, which locate in the centre tee-slot of the machine table. The width of the slot is standard and machined to close limits of size and parallelism; with these tenons in place the setting time of the machine vice is greatly reduced.

Holding the work in this fashion (figure 17.2) provides three-point contact and is therefore more satisfactory for machining, but it is important not to use a cast iron vee-block; this is particularly weak at the bottom of the vee section and is liable to fracture.

While the following method is entirely satisfactory for producing very small quantities, for larger numbers straddle milling would be more acceptable, where, after the initial setting, only the indexing of the vice through 90° is needed. The cutters would be separated on the arbor by a special collar of predetermined length.

PROCESS 17 MILLING—PRODUCING A CONCENTRIC SQUARE ON A ROUND BAR

Object To detail a method of ensuring concentricity of the square to the round bar (figure 17.1).

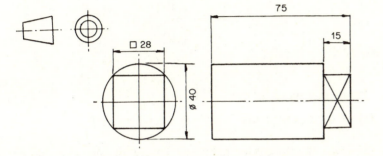

Fig 17.1

Material Bright mild steel round bar 40 mm diameter, faced both ends to 75 mm long.

Tools and equipment Horizontal milling machine fitted with arbor and sleeves, machine vice with swivel base marked in degrees, steel vee-block, side and face cutter 100×16 mm, micrometer 25–50 mm, steel rule 150 mm.

Instructions

Machine Setting

Bolt vice to machine work table, ensuring that the degree scale is set to zero or 90°.

Clamp work in vice (figure 17.2). Assemble side and face cutter to arbor and locate cutter over top face of work, raise table to obtain 'touch' of cutter on top face of work and set vertical slide indexing dial to zero.

Move work back and to one side of cutter. Raise vertical slide 15 mm and clamp.

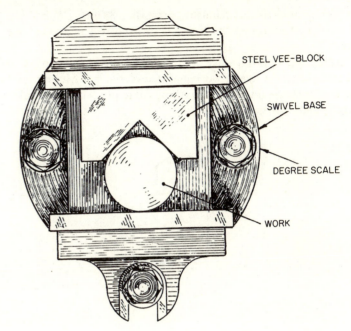

Fig 17.2

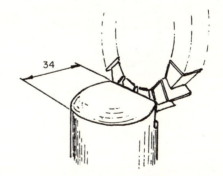

Fig 17.3

Machining

Take trial cut near position shown in figure 17.3. Measure from circumference to the trial cut face. Reset cross-slide to remove remaining stock to obtain finished dimension 34 mm. *Note* This dimension, if accurately machined, ensures correct location of one face.

Machining opposite face (figure 17.4): move table over to machine opposite side of work and take a trial cut. Measure and then index cross-slide to take finishing cut to produce two parallel flats 28 mm across.

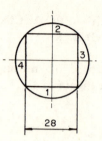

Fig 17.6

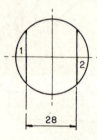

Fig 17.4

Questions: Process 17

(1)   Explain why a steel vee-block should be used.

(2)   State the reasons for and the advantages of using a vee-block in the vice.

(3)   How is concentricity ensured by following the machining instructions?

(4)   Briefly explain how the same square could be produced on a shaft 600 mm long. Show the method of work holding.

(5)   Is 28 mm across flats the largest size square that could be produced on a shaft of 40 mm diameter? Show the necessary calculations.

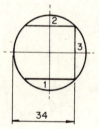

Fig 17.5

Machining to Complete Square

Swing vice 90° and secure. Locate work to cutter as in machining the first flat. Take trial and finishing cuts to obtain 34 mm dimension (figure 17.5). Complete the square as in machining the first two flats (figure 17.6). Remove burrs and sharp edges and test for concentricity.

Process 18

Milling—angular indexing

Object

Using the 40:1 dividing head for angular indexing to mill a vee-slot, for marking two lines angled to each other as shown in figure 18.3, and for converting the angle to the dividing head spindle rotation (figure 18.8).

NOTES ON PROCESS 18: ANGULAR INDEXING

A dividing head is sometimes referred to as an indexing head. Although there are many different makes obtainable, the principle on which they work is the same.

Because of the limitations of the plain *dividing head, the* universal *dividing head is much more extensively used and is selected for the following process.*

Rotation of the spindle is effected by a crank connected to a worm, which in turn drives a wormwheel. The worm is a single-start thread made of steel, hardened and ground, while the wheel is made of phosphor bronze and has 40 teeth. Hence 40 turns of the crank handle (worm) represent one turn of the spindle. This is referred to as a 40:1 ratio.

A vernier scale is attached to the main body of the dividing head to facilitate angular settings of the head spindle. For machining angular work between centres the tailstock (or footstock) is provided. This has coarse and fine adjustment for the centre height so that work angled at the head may be supported at the tailstock end.

It is vital that while indexing, rotation of the crank handle is always in one direction. If a change of direction is required then 'backlash' (clearance between worm and wheel) must be 'taken up' before reindexing. If this is overlooked errors will be encountered in the spaces indexed. Some manufacturers provide an adjustment so that clearance can be kept to a minimum and a clamping device is normally incorporated to stop clearance setting up 'chatter' during cutting.

For direct indexing it is necessary to disconnect the worm from the wheel using a quadrant mechanism built into the dividing head body. The direct indexing plate mounted behind the chuck has either a standard number of holes (usually 24) drilled in its face, or angle-sided slots machined on its circumference. The sides of the slots would need to be tapered to facilitate accurate location of the mating spring-loaded index pin.

In the following process it is necessary firstly to set the centre of the cutter central with the axis of the dividing head spindle; later it is necessary to align the cutter with the corner of the straight-sided slot that has been machined. These operations bring the student very close to the cutter. As an extra safeguard always be sure to isolate the electrical circuits of the machine before any setting operations.

PROCESS 18 MILLING—ANGULAR INDEXING

Object   To demonstrate the use of the 40:1 dividing head to mill a vee-slot and for marking-out.

Material   Flame-cut mild steel disc 12 mm thick, turned to 100 mm diameter and bored to 25 mm diameter to fit special mandrel.

Tools and equipment   Horizontal milling machine, dividing head fitted with three-jaw chuck and footstock, special mandrel, sleeve and nut (see figures 18.1 and 18.2), side and face cutter 100 × 10 mm, mandrel press, surface gauge.

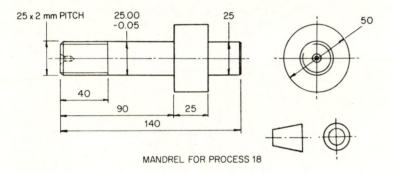

MANDREL FOR PROCESS 18

Fig 18.1

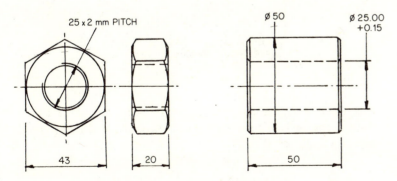

MANDREL NUT AND SLEEVE FOR PROCESS 18

Fig 18.2

Instructions

Mill the 16° vee-slot and mark the two lines shown in figure 18.3.

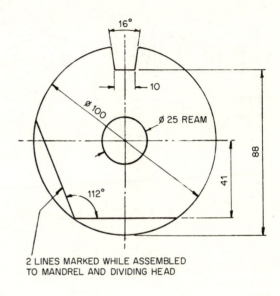

2 LINES MARKED WHILE ASSEMBLED
TO MANDREL AND DIVIDING HEAD

Fig 18.3

Preparing the Milling Machine

Fit side and face cutter to arbor, assemble dividing head and footstock to work table. Press plate on mandrel and secure with sleeve and nut. Position between dividing head and footstock as shown in figure 18.4. Locate cutter over work axis (figure 18.5). Adjust radius of dividing head crank handle so that plunger lines up with a circle of holes divisible by 9, say, a 54-hole circle. Set the index plate quadrant fingers to span 48 hole spaces (that is, $\frac{8}{9}$ of a 54-hole circle). To take up backlash between worm and worm-wheel ensure that the crank handle is turned in one direction only.

Touch work to cutter and set vertical slide to zero. Run work clear of cutter on longitudinal slide and raise table 12 mm reading on the vertical slide indexing dial.

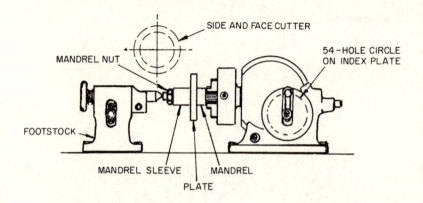

Fig 18.4

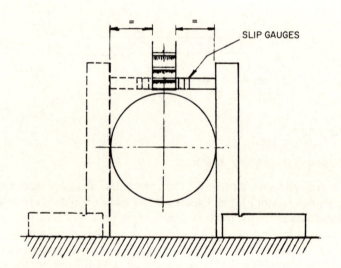

Fig 18.5

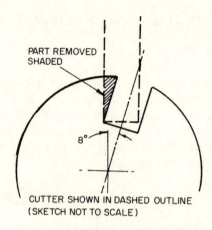

Fig 18.6

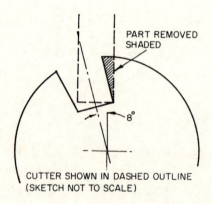

Fig 18.7

Run work back free of cutter and turn crank handle $1\frac{7}{9}$ backwards, take up backlash, realign work to cutter and mill second side of vee (figure 18.7).

Marking Two Lines at 112°

Turn crank handle to bring vee centre-line vertical. Set surface guage to 41 mm below mandrel centre and mark line across face of

Machining the Vee-slot

Mill 10 mm slot and return work to starting position. Turn crank handle 48 hole spaces, realign work to bring corner of 10 mm slot in line with corner of cutter (figure 18.6). Mill one side of vee-slot.

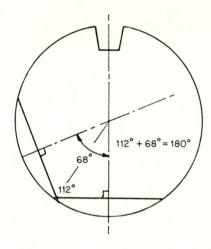

$112° + 68° = 180°$

$68°$

$112°$

Fig 18.8

footstock to the dividing head and of both to the longitudinal slide?

(5)   The machined vee-slot is of the type frequently used for direct indexing on jigs and fixtures. A matching plunger fitting closely on the tapered flanks locates the two parts. List four essential requirements of the fit between the plunger and the vee-slot.

disc. Turn crank handle $7\frac{5}{9}$ turns to rotate work 68° and mark second line (figure 18.8).

Remove sharp edges from milled slot and check vee-slot and two marked lines for obvious errors.

Questions: Process 18

(1)   Given the choice of two side and face cutters of the same diameter and width, which cutter would give the best cutting action for this particular job (a) one having 16 teeth or (b) one having 20 teeth? Give reasons.

(2)   (a) Since the vee-slot requires two opposite angular settings completed, how are the errors resulting from backlash between work and workwheel avoided? (b) Can backlash be eliminated entirely?

(3)   State what other type of milling machine can be used for this work and name a suitable cutter.

(4)   What arrangements are made for the alignment of the

Process 19

Milling a four-jaw coupling (or jaw clutch). (Figure 19.1)

Object

To study the problems of work positioning, indexing the dividing head, and providing a working clearance between the mating faces of a similar coupling.

NOTES ON PROCESS 19: JAW COUPLING

A vertical milling machine can be used to produce these couplings and to some extent have the advantage of providing the operator with a less inhibited view of the cutting process. However, it has been found that on many machine tools with the dividing head in a vertical position clearance between work and cutter is not sufficient, and for this reason a horizontal milling machine is selected.

In the first year's work the student will have learnt that rigidity is the main requirement if efficient metal-cutting, dimensional control and surface finish are to be achieved. The best set-up brings the work as close to the machine table as possible; with the dividing head in the vertical position rigidity is adversely affected. To assist in the metal-cutting process the machine slides and bearings should be adjusted with minimum clearances and the indexing slides not in use, locked.

When an odd number of teeth has to be cut the cutter can be taken clear across the blank thus finishing the side of two teeth with one pass of the cutter. When an even number of teeth is machined, as in the process sheet, it is necessary to mill all the teeth on one side and then reset the cutter for finishing the opposite side. Therefore, because of ease of production and less machining time, most clutches of this type have an odd number of teeth.

The width of cutter depends on the width of the space at the narrow ends of the teeth.

These devices are used when it is necessary to connect and disconnect the drive from two coaxial (in-line) shafts. They are sometimes known as claw-type couplings because of their interlocking teeth or projecting lugs. Because of their 'positive' engagement they transmit power without slip.

To use a jaw coupling the rotation of the drive shaft must be brought to rest and sudden starting must be acceptable.

To facilitate the drive each half of the coupling needs to be secured to its shaft, one permanently and the other allowed to move laterally on a feather key.

Common applications of this method of transmitting power can be seen on a modern centre lathe for connecting the leadscrew prior to screw cutting. Dog clutches can also be found on spring-loaded leadscrew handles of milling machines or in motorcycle gearboxes, where drive through the various ratios is via straight-sided dogs.

PROCESS 19 MILLING A FOUR-JAW COUPLING

Object   To demonstrate the problems of work positioning, indexing the dividing head and providing a working clearance between the mating faces of a similar coupling (figure 19.1).

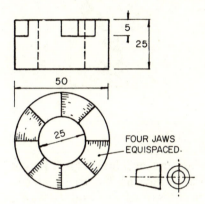

Fig 19.1

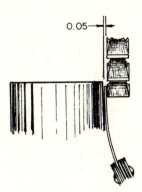

Fig 19.2

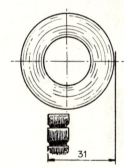

Fig 19.3

Material   A mild steel ring 50 mm outside diameter and 25 mm inside diameter faced parallel to 25 mm long.

Tools and equipment   Horizontal milling machine, dividing head, side and face cutter 100×6 mm ($4'' \times \frac{1}{4}''$), set of feeler gauges, 150 mm rule. *Note*   A horizontal milling machine is suggested for this work since it has been found that a number of vertical machines do not have sufficient clearance between cutter and work with the dividing head spindle in the vertical position.

Instructions

Setting-up

Secure dividing head to work table and swing spindle to vertical plane. Fit work in chuck to allow 8 mm to project above chuck jaws. Check for 'running true' and correct if necessary.

Fit cutter to arbor and check for cutter 'wobble'. The cutter must run true in this plane. Touch (kiss) top face of work with cutter and set vertical slide indexing dial to zero.

Run work back from cutter and raise table 5 mm to give depth of cut. Position work to side of cutter, testing touch with 0.05 mm feeler gauge (figure 19.2). Set cross-slide dial to zero.

Run work back from cutter and index cross-slide 25 mm (half work diameter) plus 6 mm (cutter width) and note *direction of feed* (figure 19.3). Put plunger of dividing head crank handle in zero hole of indexing plate. Work is now set for first cut.

Machining

Feed work through *one* side of ring (figure 19.4). Repeat to complete four cuts indexing crank ten turns (90°) each time, this will finish the right-hand flank of each jaw (figure 19.5).

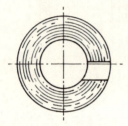

Fig 19.4

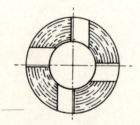

Fig 19.5

Positioning Work for Machining Left-hand Flanks

Turn dividing head crank five turns (45°). Move cross-slide back 5.9 mm (cutter width less thickness of feeler gauge 0.10 mm) (figure 19.6). *Note* It is important that the 'slack' in the cross-slide be taken up by moving past the reading and then moving forward again in the same direction as the original setting.

Proceed to take four cuts, indexing dividing head 90° after each cut. Complete left-hand flanks of each jaw.

Index work and sight 'pip' (figure 19.6) with centre of cutter.

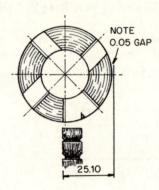

NOTE 0.05 GAP

25.10

Fig 19.6

Feed work through and machine off 'pip'. Repeat for removal of all waste material by indexing 90° after each cut.

Remove sharp edges from work and test with master coupling. With the above settings, space between jaws will be larger than jaw by 0.10 mm, thus providing a working clearance.

Questions: Process 19

(1) How is the cutter positioned to give a working clearance to the mating parts?

(2) Why is it important that the cutter should run square to the **arbor** axis (that is, 'wobble')? (See instructions.)

(3) Is cutter concentricity important? If so, explain why.

(4) In the instructions reference is made to 'slack in the cross-slide'. Explain the meaning of this.

(5) Under what circumstances would this type of coupling be used between two 'in-line' shafts?